# LUC

## *the new*

# ecological order

*translated by*
## CAROL VOLK

THE UNIVERSITY OF CHICAGO PRESS
CHICAGO & LONDON

Originally published as Luc Ferry, *Le nouvel ordre écologique:
L'arbre, l'animal et l'homme* (Paris: Bernard Grasset, 1992).
© Editions Grasset & Fasquelle, 1992.

published with the assistance of the French Ministry of
Culture and Communication.

The University of Chicago Press, Chicago 60637
The University of Chicago Press, Ltd., London
© 1995 by The University of Chicago
All rights reserved. Published 1995.
Printed in the United States of America
04 03 02 01 00 99 98                    2 3 4 5
ISBN: 0–226–24482–2 (cloth)
       0–226–24483–0 (paper)

**Library of Congress Cataloging-in-Publication Data**
Ferry, Luc.
    [Nouvel ordre écologique.   English]
    The new ecological order / Luc Ferry ;   translated by Carol Volk.
       p.   cm.
    Includes bibliographical references and index.
    1. Human ecology—Political aspects.   2. Human ecology—
Philosophy.   3. Deep ecology—Philosophy.   4. Ecofeminism.
5. Animal rights—Philosophy.   I. Title.
GF21.F4813   1995
304.2—dc20                                      94-49333
                                                  CIP

the new

# ecological order

*Our times are characterized by extraordinary intellectual romanticism:
we flee the present to take refuge in the past, any past, seeking the starry-eyed
romance of a lost security. . . . And what I wish to show is precisely that
this fearfulness is unfounded. To my mind, the European spirit is not decadent at present, but in transition; it is not in excess, but insufficiently mature.*

ROBERT MUSIL

# CONTENTS

# PREFACE

## *The Passing of the Humanist Era*

### Animal Trials

1587: The inhabitants of the village of Saint-Julien took legal action against a colony of weevils. These "creepers" having invaded the vineyards, where they caused considerable damage, the peasants called on their municipal magistrates to compose a petition in their name addressed to the "Reverend Lord Vicar-General and official of the diocese of Maurienne," whom they entreated to prescribe the appropriate measures to appease the divine anger and to undertake, "by means of excommunication or any other appropriate censor," the lawful and definitive expulsion of the tiny beasts.

Forty or so years earlier, in 1545, an identical trial had taken place against the same creepers (or at least their ancestors). The affair ended in victory for the insects, who, it is true, were defended by counsel chosen for them, as according to procedure, by the episcopal judge himself. The latter had refused to excommunicate them, arguing that as creatures of God the animals possessed the same rights as men to consume plant life; instead he prescribed numerous public prayers for the unfortunate local residents, who were required, by an ordinance of 8 May 1546, to invoke divine misericord and sincerely repent for their sins. He also invited them to pay their tithe without delay—it was the perfect occasion—and to make "three processions around the infested vineyards on three consecutive days." There followed yet further devotions or penitence of the same order. Whether due to the effect of these recommendations or, more prosaically, because of the length of the proceedings, the beetles vacated the site, and the matter ended there.

But forty-two years later, when the trial resumed on 13 April 1587, the vine growers counted on the judge's severity faced with the resurgence of the scourge. Whereupon the official merely furnished the insects with another representative or "procurator," assisted by new counsel (the one from 1545 having died in the interim). He also ordered the vicar of Saint-Julien to apply the ordonnance of 8 May 1546. This was done with great ceremony on the 20, 21, and 22 of May, as attested by the official report duly drafted and signed by the priest.

The subsequent events were every bit as complicated as those reported in the legal chronicles of today. The lawyer for the insects played so extensively and so well on the slightest points of law that on 18 July—more than three months after the opening of the trial— the accusation was beginning to waver. Sensing that the defendants' skillful arguments were likely to have a good effect (which is to say a bad one) on the official's judgment, the municipal magistrates of Saint-Julien opted for a compromise, calling a general meeting of the local residents to propose "leasing the said animals a location of sufficient pasture, outside of the disputed vineyards of Saint-Julien, from which they can draw sustenance and avoid eating or destroying the said vines." So it was: after close consideration, they decided "to offer a site named Grand-Feisse [there follows a detailed description of the plot in question], within the bounds of which lie between forty and fifty acres or thereabouts inhabited and appointed with several species of trees, plants, and foliage such as poplar, beech, cherry, oak, and plane trees, shrubs and bushes, in addition to the grass and pasture found there in fairly good quantity . . ." In short, it was a matter of convincing the adversarial party of the goodwill of the local residents and of the genuine value of the land. It should be noted that these residents did ask permission to retain a right of way to continue to exploit an ocher mine, and to take refuge on the site in case of war, but promised in so doing not to cause any damage to the "pasture of the animals" and, for good measure, "to give them a contract on the plot according to the above conditions as will be required, in due form, and valid in perpetuity . . ."

This is no doubt the first occurrence of a "natural contract," of a pact with beings of nature, but it was not enough to appease the counsel for the defense. Conscious of the weight of his accusations and determined to get the best deal for his clients, he took it upon himself to visit "Grand-Feisse," after which he concluded that, the site being "sterile and fruitless," his adversaries' claim should immediately be dismissed *cum expensis* (with costs). I do not know the official's final decision.[1] We do know, however, that other experts were called in to evaluate the true value of the land and that on 20 December the matter still was not decided . . .

By comparison with similar cases, one may assume the probable victory of the animals. It was fairly common, in fact, for the episcopal

[1] The records of this trial, which were published in 1846 by Léon Ménabréa, then advisor to the royal court of Chambéry, accurately relate its progress, including the arguments of the two counsels and the intervention of the episcopal procurator, but the outcome itself seems to have been lost. The literature devoted to animal trials is relatively limited but sometimes difficult to access. It is also, as always in cases of this sort, terribly repetitive, plagiarism being the rule. I am, therefore, obliged to say that aside from the direct sources (the records of the trial themselves, which were published in various learned journals in the early nineteenth century, and the commentaries of medieval jurists, which are invaluable resources), the two works that seemed most useful are those by Karl von Amira, *Thierstrafe und Thierprocesse* (Innsbruck, 1891), and Léon Ménabréa, "De l'origine, de la forme et de l'esprit des jugements rendus au Moyen Age contre les animaux," in *Mémoires de l'académie de Savoie* 12 (1846). There is also a near exhaustive chronological and geographical enumeration of the many animal trials known at the time in "Rapports et recherches sur les procès et jugements rendus au Moyen Age contre les animaux," by Berriat Saint-Prix, in *Mémoires et dissertations sur les antiquités nationales et étrangères,* published by the Société Royale des Antiquaires de France (1829). The article by Alexandre Sorel, "Procès contre les animaux et insectes suivis au Moyen Age dans la Picardie et le Valois," *Bulletin de la Société Historique de Compiègne* 3 (1876), contains invaluable documents, notably sentences passed. We also find an interesting analysis of the theological status of animal excommunication by H. d'Arbois de Jubainville, in the *Revue des questions historiques,* T. V. (1868). Finally, I alert the reader to the book by Jean Vartier, *Les procès de'animaux du Moyen Age à nos jours* (Paris: Hachette, 1970), which is notable for the accounts he gives of certain trials, including excerpts.

judge to take their side. The treatise *Des exorcismes* (1497), written by the Swiss theologian Felix Hemmerlein,[2] furnishes us with several such examples, including the one of the *Laubkäfer,* which for analogy's sake is worth relating here:

> Near the city of Coire, a sudden irruption occurred of larva with black heads, white bodies wide as a pinky, and six legs; they were well known to the field workers who called them, in German dialect, *Laubkäfer;* they entered the earth at the beginning of winter, plunging their murderous teeth into roots, so that when good weather returned, rather than blooming, the plants dried up . . . Now, the local residents brought these destructive insects to trial before their tribunal by means of three consecutive edicts; they provided them with counsel and a procurator in accordance with the forms of the law, then initiated proceedings against them following all the necessary formalities. Finally, the judge, considering that the said larva were creatures of God, that they had the right to live and that it would be unjust to deprive them of subsistence, banished them to a forested, untamed area, so that they would no longer have any excuse for devastating the cultivated lands. And so it was.

But it is also possible that a curse was placed on the weevils, considering what happened to the leeches of the lake of Berne in 1451. After having given the tiny beasts three days to vacate the infested waters, the bishop of Lausanne, observing that his ultimatum had gone unheeded, ventured forth in person to fulminate the following anathema: "In the name of God almighty, of the heavens, of the divine and Holy Church, I curse you, wherever you go, and you will be damned, you and your descendants, for the rest of your days on this earth."

This demonstrates that the sentence could vary depending on whether the animals were considered to be creatures of God merely

---

[2] Felix Hammerlein, 1389–1457. His two treatises were reprinted in Lyon in 1604 in volume 2 of a collection entitled *Mallei Maleficarum* (3 vols.). I am quoting here from the translation given by Ménabréa in his report.

following natural law, a scourge sent to men as punishment for their sins, or instruments of the devil opposing the ecclesiastical authority himself. In the first two cases, the imposition of penance and prayers would suffice, after which compensation might be offered to the animals, who were requested, if need be, to take up residence elsewhere; in the latter, they were "excommunicated" or, at the very least, cursed.

Hence the fact that a trial was required to decide their fate. The forms to be respected in such a case were painstakingly described by Gaspard Bally, a lawyer who practiced in Chambéry during the latter part of the seventeenth century and who was an enthusiastic partisan of these trials—which were to continue into the eighteenth century. In the second half of his work, *Traité des monitoires avec un plaidoyer contre les insectes, par spectable Gaspard Bally advocat au souverain Sénat de Savoye* (1668), he argues forcefully that "one must not underestimate monitories [that is to say, in this case, the arguments 'fulminated' by the ecclesiastical authority against the animals], seeing that they are highly important matters, carrying with them the most dangerous glaive of our Mother the Holy Church, namely excommunication, which cleaves the wood both dry and green, sparing neither the living nor the dead; it strikes not only reasonable creatures, but attaches itself to irrational ones, such as animals."[3] And Bally subsequently makes a point of indicating "how to set up one's trial in order to protect oneself from these creatures by means of the Church's malediction."

Let us reflect a moment more on this legal aspect: it is entirely indicative of a premodern, which is to say a *prehumanistic,* relationship to the animal kingdom as well as to nature in general. In most cases, the lawsuit proceeded as follows: the plaintiffs petitioned the episcopal judge, which led to a careful examination of the facts and ultimately the summoning of the animals and the assigning of a procurator (assisted, if need by, by counsel) to defend the cause of the accused.

Bally describes the following: "First, upon receiving the petition presented by the residents of the locale where damages are being

[3] This text is among the documents published by Ménabréa at the end of his report.

suffered, we obtain information as to the destruction that such animals have caused and are in danger of causing, and with this information the ecclesiastical judge assigns a curator to stand trial for these creatures, by proxy, who enumerates their arguments and defends them against the local residents who wish to force them to vacate the site they are occupying. Once the arguments are heard and considered on both sides, the judge passes sentence." Bally then gives several typical examples of petitions, of arguments put forth by local residents as well as by the counselors for the insects, models of the plaintiffs' replies, conclusions of the episcopal procurator, and finally sentences of the church judge. It is a general formula that does not take into account details or local particularities. But the picture can easily be completed in light of a few real cases drawn from the records of these trials.

Thus, for example, in each suit, aside for specifications regarding the nature and exact location of the damages, the animals incriminated—insects, reptiles, rats, mice, leeches, or others (in Marseille there was even an excommunication of dolphins, who clogged the port and made it unnavigable)—had to be described and named with great precision *so that, summoned to appear before the court, they could not claim that there had been some confusion.* This point is stressed by Barthélemy de Chassanée—a then-famous jurist who assembled in his *Conseils,* published in 1531, all that was known at the time about trials against animals. Thus we know—and this is confirmed by Felix Hemmerlein, who provides examples—that it was common to dispatch to the sites where the accused resided a sergeant or court clerk charged with loudly and clearly reading them the summons to appear, in person, on such a day, at such an hour, before the court. The summons had to be repeated three times at specified intervals, according to the custom of Roman law, for a state of nonappearance to be decreed. On the appointed day, at the appointed hour, the tribunal awaited the accused, the doors of the officiality wide open. And when, oddly enough, they did not appear, it was customary to find a plausible excuse for them in order to be able to assign a procurator— who Chassanée emphasizes held power of attorney from that point on, so long as he was not repudiated by his clients! Thus Léon Ménabréa relates, according to Thou's *L'histoire universelle* (1550), the victorious argument made by Chassanée in person on the occasion of a trial against the rats of the Autun diocese:

> While still young, he was designated to represent these animals. Although the rats had been summoned in due form, he managed to obtained that his clients be again served a writ by the priest of each parish, given, he said, that since the cause concerned all rats, all rats should be notified. Having won this point, he endeavored to show that they had not been given enough time; that it was necessary to take into account not only the distances to be traveled, but the difficulty of the journey, a difficulty made all the greater by the fact that cats were on the alert, present in every alleyway . . .[4]

During the trial of the beetles in Coire, the progress of which is related by Hemmerlein, the judge, also noting that his summons to appear for trial had gone unheeded, concluded that it was inappropriate to hold it against the little beasts "given their young age and the diminutiveness of their bodies." Once they were associated with minors, it became possible to assign them a representative, assisted by a counselor, both of whom swore to loyally serve their clients. At the trial of the leeches of Berne mentioned earlier, the bishop, not wishing to see the insects evade the court so easily, had several specimens seized in order that they be physically present at the tribunal. Once this was accomplished, he ordered that the "said leeches, both present and absent," be warned "to abandon the sites they had temporarily invaded, and to retreat to where they would be incapable of doing harm, granting them three short delays of one day each to do so, three full days in all, and this with the understanding that, once the time was up, they would risk the curse of

---

[4] Vartier contested the authenticity of this argument. It is uncertain what reasons he put forth. Whatever the case may be, it is perfectly in keeping with the spirit of these trials as well as with the recommendations given by Chassanée himself in his *Conseils* with respect to the summons to stand trial.

God and of his celestial court."[5] Finally, as evidence of the seriousness of the formal notice, the ill-fated leeches who were present in
the courtroom were executed at once!

Let us leave aside for the moment the question of the meaning
this strange theater may have had for its various protagonists. Let us
also avoid—at least for the sake of medieval historians—turning
these enigmatic practices into the truth of an age which we know today was more beautiful and more complex than the imagery inherited from the Enlightenment would lead us to believe. The fact
remains that these trials, which took place by the dozen between the
thirteenth and eighteenth centuries throughout Europe, seem unde-
niably strange to us. The problem is a classic ethnological one: how
to understand that what was a fact of life in one world can be so perfectly hermetic to another. What kind of breach must have opened
within humankind for the ritual performed in all seriousness in one
era to turn to high comedy in another?

The answer is clear to us Moderns, arising out of a concept we
take for granted: that it is insane to treat animals, beings of nature
and not of freedom, as legal subjects. We consider it self-evident
that only the latter are, so to speak, "worthy of a trial." Nature is a
dead letter for us. Literally: it no longer speaks to us for we have
long ceased—at least since Descartes—to attribute a soul to it or
to believe it inhabited by occult forces. To us the notion of crime
implies responsibility, a voluntary intention—so much so that our
legal systems grant "attenuating circumstances" in cases when the
infraction of the law was committed in an "altered state," under the
influence of our unconscious nature, thus separate from the freedom
of sovereign will. Is there truth in this or have we simply painted a
new picture which, in turn, will cause future generations to smile?
Indeed, it may well be that the separation of man and nature by
which modern humanism came to attribute a moral and legal status
to the former alone was just a brief parenthesis, marking the
boundaries of an era that is now coming to a close. Here is one
indication.

[5] Ménabréa, p. 500.

## Trees on Trial

In 1972, in the very serious *Southern California Law Review*, appeared a long article by Professor Christopher D. Stone entitled: "Should Trees Have Standing? Toward Legal Rights for Natural Objects." Republished two years later in the form of a short book, Stone's article experienced great success in a context that is worth relating here. Though it seems light years away from our medieval countryside, contemporary California has nonetheless attempted to reinvent the idea of a law of natural beings, in the course of what turned out to be an extraordinary trial.

In 1970, the United States Forest Service granted Walt Disney Enterprises a permit authorizing them to "develop" a wild valley, Mineral King, situated in the Sierra Nevada. A budget of thirty-five million dollars was planned for the construction of hotels, restaurants, and play areas, based on the model of Disneyland. The powerful Sierra Club, probably one of the most capable ecological associations in the world, filed suit, alleging that the project threatened to destroy the aesthetic and natural equilibrium of Mineral King. The suit was rejected by the court, not on the grounds that the Forest Service was right to issue the permit but on the grounds that the Sierra Club had no claims to support the plea, since its interests were not *directly* encroached upon by the project in question (let us not forget that American law rests in principle on the idea that the legal system as a whole exists to protect *interests*, whatever they may be, and not abstract values). When the affair moved into appeals, Professor Stone, who until then had been calmly defending the ideas of radical ecology in his university courses, set about to rapidly draft an article proposing, in his own words, "that we give legal rights to forests, oceans, rivers and other so-called natural objects in the environment—indeed to the natural environment as a whole." He had to act quickly so that the judges would have a precedent at their disposal, albeit a theoretical one. As Stone writes in the preface to his book, "Perhaps the injury to the Sierra Club was tenuous, but the injury to Mineral King—the park itself—wasn't. If I could get the courts thinking about the park itself as a jural person—the way corporations are 'persons'—the notion of nature having rights would here make a significant operational difference . . ." Conclusion: of the nine judges, four voted against Stone's argument, two abstained, *but three voted for it,* so that it can be said the trees lost their trial by one vote . . .

Stone's argument in favor of the rights of objects is not without interest. Its first point, which should delight disciples of Tocqueville, consists in recalling the reasoning—which is standard within this ecologist literature—according to which the day of the rights of nature has now come, after that of children, women, blacks, Indians, even of prisoners, the insane, or embryos (within the context of medical research, not to mention abortion legislation . . .). In short, Stone suggests that what seemed "unthinkable" at one time, often not very long ago, has now become perfectly acceptable. And he cites to felicitous effect the judgments of a certain court of law which, as late as the nineteenth century, considered that, to varying degrees, Chinese, women, and blacks could not hold legal rights.

The requirements for declaring a being a "bearer of legal rights" have naturally yet to be defined. According to Stone, it is necessary, *first,* that this being be able to bring legal action on its own behalf; *second,* that in an eventual trial the court be able to consider the idea of damages or harm brought to this being (and not, for example, to the being's owner); and *third,* and last, that the eventual compensation benefit this being directly. The rest of the work is devoted to showing, point by point, that trees (and other natural beings) can easily satisfy these three conditions, provided, of course, that we accept, as we do in other comparable cases, for other nonreasoning entities, that the subject take legal action by intermediary of its representatives (ecological associations or others); Stone goes so far as to envisage a proportional representation for trees on the legislative level! An analogous thesis is now being adopted in France by a certain number of jurists who are also basing themselves on the principle that the tradition of modern humanism, according to which only humankind has legal standing, should be questioned. Marie-Angèle Hermitte, for instance, looks favorably upon the few precedents by which one "turns a zone, chosen as a function of its interest as an ecosystem, into a legal subject, represented by a committee or an association responsible for asserting that subject's right to be itself, which is to say its right to remain in the state in which it stands or to return to a superior state."[6]

☙

Make no mistake about it: these eminent jurists are sane. From a pragmatic or operational standpoint, Stone's argument, even if it can be contested, as we shall see later, is not without coherence: such an argument would make it possible to bring suits against large polluters de facto, in the absence of a direct interest (Stone cites the concrete and indeed problematic case of various enterprises that devastate the environment, yet cannot be stopped because the pollution affects zones where no immediate individual interest is being encroached upon). On a quasi "ontological" level, however, the questions become more pressing—the astute legal construction conceals a questionable philosophical bias in favor of a return to former conceptions of nature. For can it not be said that these thinkers, who claim to be "postmodern" in the literal sense of the word—philosophers of jurists of "posthumanism"—are in communion with a *premodern* vision of the world, *in which beings of nature recover their status as legal subjects?* Is it not also strange to us, insofar as we are still Moderns, that trees or insects can win or lose a trial?

Thus the humanist era is being brought to a close; and this is the main objective for these new zealots of nature. Its oddities aside— and we would be wrong to think they escape Stone and his friends— the debate on the rights of trees, islands, or rocks is based on no other grounds: it is a matter of determining whether the only legal subject is man, or whether, on the contrary, legal status should extend to what is today called the "biosphere" or the "ecosphere," formerly known as the "cosmos." From every point of view—ethical, legal, or ontological—man would be but one element among others, and the least *sympathetic* one at that, being the least *symbiotic* with the harmonious and orderly universe into which he is constantly, by his excess,

[6]Marie-Angèle Hermitte, See "Le concept de diversité biologique et la création d'un status de la nature," in *L'homme, la nature, le droit* (Bourgeois, 1988).

by his *"hubris,"* introducing the worst disorder. Is it not time for a new "natural contract" to check this egoism and reestablish the harmony that has been lost? Is it not in this direction, from a humanistic vision of law to a cosmic one, that this premodern postmodernity invites us to advance?

## The Opiate of the People or the New Ideal

The new cosmology emerging from these trials, in which trees are elevated to the status of legal entities, is seductive in more than one way to those disappointed by the modern world, which is to say all of us to varying degrees. The truth is it has a bit of everything, or almost, even the most classical elements of the now-defunct "great political plans." Set against the idea of a cosmic order, ecology—*this* form of ecology, that is, for we shall see that there are others—reconnects with the notion of "systems," which we thought thoroughly discredited. It is at this price—which may seem too great—that it can call itself a true "world vision," whereas the decline of political utopias, but also the parcelization of knowledge and the growing "jargonization" of individual scientific disciplines, seemed to forever prohibit any plan for the globalization of thought. This systemic, if not systematic, pretension is indispensable to the foundation of a political eschatology. At a time when ethical guide marks are more than ever floating and undetermined, it allows the unhoped-for promise of rootedness to form, an objective rootedness, certain of a new moral ideal: purity recovers its standing, but it is no longer founded on a religious or "ideological" belief. Instead it claims to be "proven," "demonstrated" by the incontestable facts of a new science—ecology—which, though global, as was philosophy, is nonetheless as beyond question as the positive sciences on which it bases itself. If the health department has shown that smoking causes serious illness, if laboratories have determined the disastrous effects of aerosols, if automobile makers themselves are forced to recognize a connection between exhaust fumes and deforestation, isn't it senseless, even *immoral,* to continue along the path of depredation? And is it not the modern world as a whole—Stone is right to insist—with its arrogant anthropocentrism in industry as in culture (are the two still separate?), that should be incriminated?

While strong political ideologies, with the exception of religious fundamentalisms, are in decline the world over, is there not something here to revive the eternal flames of militantism? Especially since the critique of modernity can count not only on the fervent support of the major religious, which are always quick to reprove the vanity of men, but also on the approbation of neofascists or ex-Stalinists, who, with their antiliberal convictions, past or present, repressed out of necessity more than out of reason, are only too glad to embark on a new adventure in science-based politics.

"Ecology or barbarism": this may well be the slogan of the next century. It is, therefore, important to distinguish the false debate that threatens to emerge and the real question still waiting to be addressed.

The false debate is simple and already familiar to us: the "vigilant democrat" calls the ecologist a "fascist" on the grounds that his love of nature is too redolent of the fatherland not to be a bit khaki in its green. Another variation: the same vigilant democrat detects a reincarnation of leftism in the critique of Western civilization and the praise of the frugal life led by, say, the American Indians. Let us be clear: the democrat is not entirely wrong, far from it. He is right to encourage us to reflect on the two perverse tendencies of contemporary ecologism, both driven by the same disdain for formal social democracy, both connected to a solid tradition that had its peak some time in the late 1930s. But ultimately, we cannot boil the challenges posed by ecology to the tradition of modern humanism down to nothing—to the mere fantasies of alarmist political ideologies. Especially since the "average" ecologist sensibility, that of the man on the street, has nothing extremist or antidemocratic about it, but derives more from an ethics of authenticity, from a concern for the self, in the name of which one insists—and why not?—on a certain "quality of life."

And herein lies the real question. Our entire democratic culture, our entire economic, industrial, intellectual, and artistic history since the French Revolution has been marked, for basic philosophical reasons, by the glorification of *uprootedness,* or *innovation,* which amounts to the same thing—a glorification which romanticism, followed by fascism and Nazism, have continually denounced as ruinous to national identity, even to local particularities and customs. The antihumanism of these movements, which was explicit on a cultural level, was accompanied by a concern for rootedness that lent itself to the development of a great attraction to ecology. To parody Marcel Gauchet's felicitous phrase, "the love of nature" (poorly) concealed "the hatred of men."[7]

It is not by chance, then, that the Nazi regime, and Hitler personally, are responsible for the two most detailed legislations regarding the protection of nature and animals in the history of humanity.[8] And yet, we cannot deny that the "hatred of men," understood in another sense, as the Cartesian disdain for nature, and particularly for living beings, is also a very real question. We cannot help but recognize that *metaphysical* humanism was essentially at the origin of an unprecedented colonization of nature—whether we take this to mean territories or living beings, animals or "naturals," as the "indigenous "used to be called.

Is a nontyrannical, nonmetaphysical humanism possible? Would it have something other to say than Cartesianism, so concerned with making man the "master and possessor of nature"—or is the only solution to get "down to earth," to return to old-time frugality, to the *wilderness* in which American cinema and German philosophy constantly immerse us? Would such a move signal the end of all that we may love about modern culture, artificial and unnatural though it may be? The question here is whether the civilization of uprootedness and innovation is utterly irreconcilable with a concern for nature, as appears, *initially,* to be the case. And, conversely, whether the latter implies a renunciation of artifice. I do not believe so. All the same, if we wish to outline the conditions for a reconciliation, we must realize that we can no longer speak of ecology in the singular. The philosophies that implicitly or explicitly underlie the various sensibilities on questions of the environment are so varied, even so opposed to one another, that no one statement applies to all. The time has come to take stock of this complexity.

[7] See Marcel Gauchet, *Le Débat,* no. 60 (August 1990).
[8] See in Part Two of this book the chapter on "Nazi Ecology."

## The Three Ecologies

In France, home of Descartes, but also in most of the Catholic countries of Southern Europe, ecology has yet to find theoreticians comparable to those of the Anglo-Saxon or Germanic world. The reasons for this are unclear; the hypothesis that proposes a link between religion and the concern for nature no doubt merits further investigation. Generally, it may be observed that wherever theoretical debates on ecology have taken coherent philosophical form they have been structured into three currents that are distinct from or even entirely opposed to one another with respect to the seminal question: that of the relationship between man and nature.

The first is no doubt the most ordinary, but it is also the least doctrinaire and, therefore, the least dogmatic; it is based on the idea that, by protecting nature, man is still first and foremost protecting himself, even if it is from himself in his capacity as mad scientist. The environment is endowed with no intrinsic value here. Rather this scenario stems from an awareness that by destroying the milieu that surrounds him, man may be endangering his own existence or, at the very least, depriving himself of the conditions for a good life on this earth. Thus nature is taken only *indirectly* into consideration, based on a position that may be classified as "humanist," even *anthropocentrist;* it is considered merely to be the human environment, literally that which surrounds him—the periphery, then, and not the center. As such, it cannot be considered a legal subject, an entity possessing absolute value in and of itself.

The second current takes a step in the direction of attributing moral significance to certain nonhuman beings. It consists in giving serious consideration to the "utilitarian" principle according to which one must not only look out for man's best interests, but, more generally, try to both diminish the total suffering in the world as much as possible and increase the quantity of well-being. From this perspective, which is quite common in the Anglo-Saxon world, where it is the basis for the enormous animal liberation movement, all beings capable of feeling pleasure and pain must be considered legal subjects and treated as such. The anthropocentrist point of view is thus discredited within this frame work, since animals are included, by the same token as men, within the sphere of moral considerations.

The third tendency is the one we have seen at work in the call for the rights of trees, which is to say of nature in and of itself, including in its vegetable and mineral forms. Let us beware of dismissing this current too quickly. Not only is it tending to become the dominant ideology of "alternative" movements in Germany and the United States, it is also the one that raises the matter of the need to throw humanism into question in the most radical terms. It has, of course, found its own intellectuals. Included among these are Aldo Leopold in the United States, Hans Jonas in Germany, whose 1979 *The Imperative of Responsibility {Das Prinzip Verantwortung}* has sold over a hundred and fifty thousand copies and become the bible of a certain German Left and beyond, and Michel Serres, author of *Le contrat naturel {The Natural Contract},* whose theses are probably not truly understood in France for what they are: an authentic American-style crusade against anthropocentrism (Serres has been teaching in California for many years and is well-acquainted with this literature) in the name of the rights of nature. For this is the primary issue in this third version of ecology—that the old "social contract" devised by political thinkers must give way to a "natural contract," in which the entire universe becomes a subject of law: it is no longer a matter of defending man, considered as the center of the world, from himself, but rather of defending the *cosmos* from him. The ecosystem or "biosphere" is endowed with an intrinsic value far superior to that of this species—this generally quite destructive species that is the human race.

According to a terminology now classic in American universities, a distinction must be made between "deep ecology," which is "ecocentric" or "biocentric," and "shallow" or "environmentalist ecology," which is based on the old anthropocentrism. For more than twenty years now, without creating the slightest ripple in France before the publication of the book by Serres (who remains highly discreet as to his sources), a wealth of literature has been contributing to the construction of a coherent doctrine of nature as a new legal subject. It is now necessary to take stock of these developments.

But it is appropriate to consider the tensions that make ecology movements so complex from yet another perspective. *For the renais-*

*sance of feelings of compassion for natural beings is always accompanied by a critique of modernity*—designated, depending on the frame of reference, as "capitalist," "Western," "technological," or, more generally, "consumerist." The different critiques of the modern world can take highly varied forms, thus offering a baseline for a new typology of the faces of ecology.

## The "Thirties" or The Three Critiques of Modernity

We should begin by stating a paradox: the concern for the environment is most pronounced, it seems, within liberal-social-democratic societies. Even in France, where the love of nature is reputed to be less keen than in Germany or the United States, the defenders of the sea and land have many sympathizers: Jacques Cousteau is always listed among the personalities most dear to the hearts of French citizens, and philosophers, whose scholarly or academic works used to be private affairs, have become popular thinkers, exposing the principles governing ecosystems or popularizing the Anglo-Saxon idea of a "natural contract" giving long overdue legal status to nonhuman beings. True, there have been punks rebellious enough against the "goody-goody" young pacifists to sport "Nuke the Whales!" buttons. But they wear them as a joke, precisely to emphasize a fringe status. For among Westerners, ecology is like the old adage that says that one can never be too rich or too thin: no one can seriously deny that it has a certain legitimacy.

In the Third World or in the countries of the East, the necessities of economic development relegate environmental questions to secondary status. Herein lies an enigma. For, far from being dominated by the mere logic of capitalist profitability or blinded by the science-minded ideology which supposedly governs the world of technology, our liberal democracies beget their own critiques, including highly radical ones. It is in the West that the ecologist denunciation of Western wrongdoings gains the most acceptance, that the most sophisticated arguments are developed and the most sympathizers engaged. And this paradox is all the more striking in the case of the radical ecology movement, which launches the most negative critiques ever pronounced against the modern world: Nazism itself, not to mention Stalinism, maintained an ambiguous attitude in the face of technoscience, denouncing it on the one hand yet constantly developing it on the other, in the bellicose context of a "total mobilization." How is this strange phenomenon to be understood?

A first possibility, that of the ecologists most hostile to Western civilization, consists in voluntarily marginalizing the protest: to highlight its subversive aspects and rechannel revolutionary myths, protestors call for the death of the "system," as if they themselves hailed from its margins. On the classic model of the bourgeois student denouncing the activities of the bourgeoisie, or the media intellectual "courageously" scolding the media each time a platform is opened to him, the radical ecologist convinces himself that his struggle breaks with the universe he wishes to destroy. The fact that the most industrialized societies are also those in which his argument is best received doesn't trouble him. On the contrary, he sees this as confirmation of his exteriority: it is precisely because he lives in one of these destructive countries that awareness can come to those who are courageous enough to reflect on these issues.

Another possibility, a less romantic but more plausible one, consists in considering that the demand for a healthy environment, in which the *well-being* of all living creatures is ensured, is related to the call for a Welfare State, the blossoming of which remains, without question, a particularity of Western culture. From this perspective, the attention brought to nature would not be forged so much against the modern universe as *produced by it,* resulting, basically, from the same democratic passions that prompt demands for a right to leisure, health, and so on, demands eminently characteristic of the modern relationship between the individual and a liberal State turned guardian.

We sense here the extent to which the critiques of modernity emanating from ecology movements are apt to be in opposition. First of all, it is possible to denounce the real or imagined misdeeds of the liberal world in the name of *nostalgia,* or, on the contrary, in that of *hope:* either the romantic nostalgia for a lost past, for a national identity flouted by the culture of rootlessness, or revolutionary hope in a radiant future, in a classless and free society. Despite their inherent differences, fascism and communism thus would share the same wariness of formal democracy, the same repugnance toward the market and the plutocratic society it naturally engenders, the same concern with producing a new man, the same myth, essentially, of uncompromised and uncompromising purity. In both cases, the critique of modernity would profess to be *external,* pronounced in the name of a radically different alternative world, whether premodern or postmodern. For nothing is to be retained from liberal free-market politics, no concessions to be made. The only appropriate political attitude, faced with this basic evil, is revolution, whether neoconservative or proletarian, not reform.

This thirties-style pathos finds a new home in deep ecology. As opposed to its "superficial" competitor, which it considers blatantly reformist, it adopts a "radical" attitude: no compromise is possible with the Western way of life, with "Western Civilization." The West is not "politically correct." Not only has it failed, but in its fall it also brings down the peoples of the Third World, ethnic minorities and dominated factions—whether women, or anyone who is "different" in any way. Only those who fall to the extremes are acceptable.

Hence the fact that deep ecology can combine in one movement both the traditional themes of the extreme Right and the futurist ideas of the extreme Left. The essential element which gives coherence to the whole is the heart of the diagnosis: anthropocentrist modernity is a total disaster. Opposing the tendency toward one-dimensionality, already described by Marcuse or Foucault, opposing the "political-media lobby," uniformity, consensus, and claims toward universality, deep ecology praises diversity, singularity, particularity—in other words, both that which is "local" (leftist version of deep ecology) and that which is "national" (right-wing version). The entire problem, of course, is that the models of reference, fascism and communism, having crumbled, this external critique is desperately seeking conceptual markers that will allow it to move beyond knee-jerk responses. It is therefore necessary to understand how the two totalitarianisms, which in other times would have constituted the opposite poles of an ideal, appear only spottily, reduced to the status of inclinations or *intentionalities.* Yet the two fundamental tendencies of this type of ecology, its two possible interpretations, can nonetheless be distinguished by their relentless hatred toward all forms of humanistic culture — in particular the disgraced heritage of the Enlightenment.

The third form of ecology is another case entirely, coinciding in large part with that designated as "environmentalist." It is true that it too is based on a critique of modernity, but this time the critique endeavors to be *internal,* thus reformist. Propelled by highly democratic passions such as a concern for the self, respect for the individual, the quest for a more "authentic" existence, for a better, less stressful quality of life, in which a dose of solitude provides an antidote to big-city crowds, it aspires more to overhaul the system than to replace it. Moreover, though a lover of deserted beaches and unpolluted seas, the democrat ecologist would be hard-pressed to forgo the benefits of modern science and the company of others. It is difficult to imagine him renouncing the progress of medicine for his children or for himself, or even the advantages of a private car or air travel. He has no taste for extremist political solutions, and collective authoritarian decisions do not sit well with his respect for individual autonomy. The deep ecologist may accuse him of incoherence . . .

What can he respond?

Quite a few things, actually. Starting with the fact that the hatred of the *artifice* connected with our civilization of rootlessness is also a *hatred of humans as such.* For man is the antinatural being par excellence. This is even what distinguishes him from other beings, including those who seem the closest to him: animals. This is how he escapes natural cycles, how he attains the realm of culture, and the sphere of morality, which presupposes living in accordance with laws and not just with nature. It is because humankind is not bound to instinct, to biological processes alone, that it possesses a history, that generations follow one another but do not necessarily resemble each other—while the animal kingdom observes perfect continuity.

As strange as it may seem in the land of Descartes, the animal kingdom thus turns out to be at the heart of contemporary debates on the relationship between man and nature. There are at least three reasons for this. The first is due to the fact that the love of animals has become a mass phenomenon common to most democratic societies: there are thirty-five million pets in France alone! The second reason why the antimodern themes of ecology converge on the animal kingdom is more philosophical: Cartesian humanism is without a doubt the doctrine that went the farthest in devalorizing nature in general and animals in particular. Reduced to simple mechanistic states, they have been denied intelligence, affectivity, and even sensitivity. The theory of the animal-machine is the quintessence of what a certain contemporary ecology denounces under the name of anthropocentrism. The animal, then, is the first being one encounters in the process of *decentering*, which leads from the questioning of anthropocentrism to the adoption of nature as a legal subject. In passing from man to the universe, as not only deep ecology but utilitarianism requires, one passes first by the animal.

If the question that guides this essay is that of the capacities of a nonmetaphysical humanism to take responsibility for matters of the environment, it is by the particular, but paradigmatic, case of animals that the discussion must begin.

*part one*

ANIMALS, OR
THE CONFUSION
OF GENRES

# Antinatural Man

Aristotle or Descartes? I sometimes wonder which of the two would meet with greater disfavor in the eyes of our contemporaries should they happen to take it upon themselves to read them. The first for justifying slavery as "natural," or the second for making such a sharp distinction between men and animals that he came to consider the latter as simple machines? The love of living creatures has made great progress in France, land of *The Discourse on Method* and *Metaphysical Meditations!* Today, for instance, we can celebrate the existence of a "French League for Animal Rights," founded by Alfred Kastler, Etienne Wolff, and Rémy Chauvin. And even a "Universal Declaration of Animal Rights," the result of deliberations led by renowned jurists and scientists, was created in 1978, barely thirty years after René Cassin's "Universal Declaration of Human Rights." In it, we read that "all animals are born equal and have the same rights to exist." We also learn that "genocide is being perpetrated by man," even though, as part of the "animal species," he cannot "exterminate other animals" without violating their most indefeasible rights.

Do words still have meaning? In the zoophile spirit that impregnates our democratic culture, the ideas that a distinction between humankind and animal-kind may possess ethical significance seems intolerant, an indication of a spirit of segregation, of exclusion even, at a time when the right-to-be-different ideology reigns almost exclusively. Indeed, doesn't science teach us that a secret continuity exists between living creatures? In the name of science, then, it is proper to grant equal respect to all manifestations of life in the universe. A *sympathetic* goal, to be sure, but perhaps one that is incompatible with the terms by which the secular humanism that emerged

3

from the French Revolution has defined itself. We may not like this inheritance, we may even wish to "deconstruct" it, to finish it off once and for all. Still we must evaluate the implications of such a rupture—which naturally presupposes discovering what they are.

To this end I suggest suspending our sympathy for a moment, if only provisionally, and taking a fresh look at the manner in which an anthropology was formed during the century of the Enlightenment without which the ethical universe derived from the Revolution would be deprived of its deepest philosophical dimension. Here we have a paradox: the key is found in Rousseau, who, as we also know, was one of the great innovators of the romantic sensibility. He was the first to draw the consequences of the Cartesian distinction between animals and men for the emergence of a world of distinctively human culture (or as Dilthey would have said, a "world of the mind"). In one of those brilliantly illuminating passages, which in a few phrases manage to express a thought whose age-old repercussions are still with us, the *Discourse on the Origin and Foundations of Inequality* elaborates a reflection on humanity without which our intellectual universe would not be what it is today. By that I don't mean to say that this page was to play a determining role in our history. But at the very least it sets the tone for a crucial moment. Hegel said that philosophy was "capturing one's time in thought." Rousseau's text illustrates, perhaps as never before, the pertinence of this statement. Early on, he remarks the following:

> In every animal I see only an ingenious machine to which nature has given senses in order to revitalize itself and guarantee itself, to a certain point, from all that tends to destroy or upset it. I perceive precisely the same things in the human machine, with the difference that nature alone does everything in the operations of a beast, whereas man contributes to his operations by being a free agent. The former chooses or rejects by instinct and the latter by an act of freedom, so that a beast cannot deviate from the rule that is prescribed to it even when it would be advantageous for it to do so, and a man deviates from it often to his detriment. Thus a pigeon would die of hunger near a basin filled with

the best meats, and a cat upon heaps of fruits or grain,
although each could very well nourish itself on the food it
disdains if it made up its mind to try some. Thus dissolute
men abandon themselves to excesses which cause them
fever and death, because the mind depraves the senses and
because the will still speaks when nature is silent.[1]

The phrase that closes the evocation of the misery which only
freedom can bring to man adds beauty to the depth of the observation. Like the flowering bulb, it contains unsuspected multiplicity.
On the surface, we are looking at an opposition between nature and
freedom. This means, first of all, that the animal is programmed by a
code which goes by the name "instinct." Whether a grain- or a meat-
eater, it cannot liberate itself from the natural rule controlling its behavior. Determinism is so strong it can even bring death, where an
infinitesimal dose of freedom with respect to the norm would allow
for easy survival. The situation of the human being is the opposite.
He is indetermination par excellence: he is so oblivious to nature it
can cost him his life. Man is free enough to die of freedom, and this
freedom, as opposed to what the Ancients thought, is potentially
harmful. *Optima vide, deteriora sequor.* Seeing the best, he can choose
the worst: this is the motto of the antinatural creature. His *humanitas*
resides in his freedom, in the fact that he is undefined, that his nature
is to have no nature but to possess the capacity to distance himself
from any code within which one may seek to imprison him. In other
words: his essence is that he has no essence. Romantic racialism and
historicism are thus inherently impossible.

For what is racism at its philosophical core if not the attempt to
define a category of humans by its essence? There would be a black
essence, an Arab essence, and a Jewish essence, as a function of which
each individual would possess certain insurmountable characteristics inscribed in his or her makeup. The fact that we might then
seek to classify defects and qualities on a hierarchical scale by race
then becomes, if not irrelevant, at least *secondary.* Inequality is a

[1] From Rousseau, *The First and Second Discourse,* translated by Roger D.
and Judith R. Masters (New York: St. Martin's Press, 1964).

by-product, but the damage is done as soon as we posit the existence of definitions. Philosemitism is just as suspect as antisemitism, its opposite. What Rousseau affirms is the possibility for man as man, for this "abstract" man of which all counterrevolutionary thought would contest the existence, to separate himself from biological determinations, from what Hannah Arendt would later call "the life cycle." The question of history is analogous to that of race. Abstract humanism contains within it the idea that "history," according to Rabaut Saint-Etienne's felicitous expression, "is not our code." Contrary to what romanticism would later develop, man is no more prisoner of his linguistic or national tradition than of his biological being. This, in fact, is why revolution is possible; it is the supreme act of the free man, extricating himself from the universe in which he was supposed to be shaped and contained. To learn a foreign language, to open oneself up in this way to understanding another culture, remains and will always remain, in this sense, the mark of a "broadened mind."

No doubt one will object that such a vision of things is "idealistic," that it discounts the teachings of biology and sociology—which continue to maintain that the individual is produced, determined, and molded by his membership in a social or biological group. Biologically, the objection extends in two directions: humans bear animal traits, animals bear human traits. In support of the latter, for example, one can cite the suicide of whales—an indication that they too can distance themselves from their natural tendency—the language of monkeys and of dolphins, the capacity of certain animals to manipulate tools in order to realize their objectives, not to mention canine devotion or feline independence . . . The problem, of course, is that this separation from the commandments of nature is not transmitted from *one generation to the next* to form a history. A separation from natural norms only becomes evident when it engenders a cultural universe, and animal societies, as I'm sure we'll all agree, have no history.

Sociologically, on the other hand, one would have to insist that the difference is only a matter of degree, and this would serve to explain Rousseau's error: it is because the number of brain cells is very high in man that we observe a moment of indetermination between a stimulus and its response. The animal brain being less complex, the natural determination—the "instinct" of which Rousseau speaks— is more apparent. The truth is that, from the point of view of the scientist who observes "neuronal" man without the philosophical prejudices of another era, this indetermination is an illusion: we cannot use it to justify the hypothesis of a free will. According to determinist thought, freedom is nothing but the "madness" which Spinoza already showed is linked to our ignorance of the subtle material mechanisms surrounding us: greater knowledge would allow us to credit to nature what we naively attribute to freedom.

Based on the work of biologist Henri Laborit, Alain Resnais chose to show in *My American Uncle,* one of his strange thesis films, how our behavior in daily life is no different from that of laboratory rats: stress makes us aggressive, reward or failure prompts us to pursue or abandon our undertakings, such that, in most cases, we are unable to control our emotional reactions. Ultimately our supposed affective or intellectual "choices" are reduced to unconscious reflexes. Under such conditions, it is indeed difficult to see why the protection afforded us by the rights of man should be reserved only for beings that an arbitrary decree qualifies as "human." Or rather, if we admit that the value of a creature is a function of its biological complexity, would it not be appropriate to abolish all clear legal distinctions in favor of a graduated concept of rights?

From the point of view of sociology, the objections are analogous: determined by our biological nature, we are prisoners of our social conditions, in particular those inherent to our class. The story is too well known to insist on it here. The response, too, is well known, which consists in recalling that we cannot combine a "situation" and a "determination" in a single concept: it is clear that we have a body, that we live within a social milieu, a nation, a culture, and a language. I can't jump as high as a tiger. Does that mean that I am less free than he?

The affirmation of freedom does not mean that one must deny factual evidence. It requires only that it be possible for man to escape reification, to emancipate himself from that which continually threatens to transform him into a thing. It postulates a margin for maneuvering, the ability to avoid having what is on the order of a simple situation transform itself into a determination (which of course is often a very real possibility). Every individual is in some part reified, and sociology—and by the same token psychoanalysis—may attempt to account for this. Worse, we have every reason to fear that reification will be the ultimate horizon of our lives—which poses the difficult problem of aging in a democratic universe that has raised the heroism of freedom to cult levels. But if we assert that no one can rise above his condition we sink into a form of "classism," and one is hard-pressed to see how this classism is preferable to ordinary racism or sexism. Naturally it is not a matter of proving the existence of transcendence. But by denying it, we essentially misjudge man's aptitude for creating a universe that breaks with the natural (unconscious) dimension of social or biological life. This is what the second portion of Rousseau's text suggests:

> But if the difficulties surrounding all these questions should leave some room for dispute on this difference between man and animal, there is another very specific quality that distinguishes them and about which there can be no dispute: the faculty of self-perfection, a faculty which, with the aid of circumstances, successively develops all the others, and resides among us as much in the species as in the individual. By contrast an animal is at the end of a few months what it will be all its life; and its species is at the end of a thousand years what it was the first year of that thousand. Why is man alone subject to becoming imbecile? Is it not that he thereby returns to his primitive state; and that— while the beast, which has acquired nothing and which has, moreover, nothing to lose, always retains its instinct—man, losing again by old age or other accidents all that his *perfectibility* has made him acquire, thus falls back lower than the beast itself?

With rare expressive density, this passage announces what the modern universe presents perhaps at its deepest, but also at its scariest, on three levels: anthropological, ethical, and existential. Here again, we have to consider the implications. We should first note the direct link uniting the definition of man to the emergence of a modern problematics of historicity. Kant, Fichte, and Sartre were to return to this theme: it is because man is originally "nothingness," because, as opposed to animals, he is *nothing* as determined *by nature,* that he is devoted to a history which is a history of freedom. The "perfectibility" of which Rousseau speaks will thus develop on two levels: *education understood as the history of the individual, and politics as the history of the species.* Rousseau's passion for the land of childhood, which it has been said he discovered, is equaled only by his interest in politics. *Emile* and *The Social Contract.* In both cases, the problems are analogous: it is a matter of knowing how a being whose essence it is to transcend all individual determination can create himself over time without becoming *something* and thus losing himself in reification (what Sartre also called "bad faith"). The parallel suggested here between education and politics will remain a constant of modern thought, from Kant until Marx, the theoretician of praxis. The birth of "active methods" of pedagogy as well as that of political militantism are inseparable from a vision of man as transcendence or as nothingness—as a being who cannot be reduced to the situations in which he finds himself perpetually ensnared.

Here we see all that separates such a vision of historicity from romanticism: civilization cannot be reduced to the national, linguistic, or cultural traditions to which one *automatically* belongs (unconsciously and involuntarily); far from being bound exclusively to the values of *rootedness,* it finds its true elevation in this separation from the natural universe by which "a world of the mind" is progressively forged.

One may object, and not without reason, that no man can claim to detach himself totally from the historical and cultural community to which he belongs, that abstract universalism is neither credible nor even desirable: How can we help but be attached in some way to what the romantics called *Heimatsgefühl,* to this place, to this time, to this tradition in which we ultimately feel at home (*bei sich*)? This objection cannot be easily dismissed. It therefore needs to be clarified.

In the conflict between romanticism and the Enlightenment, two concepts of culture and history are at odds, each with solid arguments against the other. According to the former, a man can only be a true man when he is among his own kind, in the community

that surrounds him and fashions him on the model of its language—
a language we learn but don't create ourselves. Hence the critique
that is constantly addressed to the partisans of freedom conceived as
transcendence: when deprived of his culture and cut off from his
roots, the man who aspires to freedom in reality loses what makes
him human. If possessing a culture is truly what distinguishes man
from animals, by emancipating himself from this culture he essentially joins the ranks of nonhumans.

It is in this sense that being uprooted is dehumanizing. The
Enlightenment philosopher—and *on this point* Rousseau is one representative, the one that French revolutionaries would follow—is also
of the opinion that culture and history are particular to man. But
in his eyes, romanticism *naturalizes* this particular characteristic. It
turns it into a second nature, so to speak, comparable to that imposed
upon men *from without* and determining them so completely that, in
trying to escape it, they tip into the void. For him, history is not *tradition,* it is *creation, innovation, perfectibility.* It is not received by men
from without but is constructed by them; it is not the negation of
their liberty in the name of an intangible past but rather the effect
of this liberty inscribed in a dynamic of the future. To the romantic
who holds that the abstract man is no longer a man, the *Aufklärer*
responds, on the contrary, that it is the rooted individual, wholly
determined by his situation, who returns to nature and thereby loses
his human quality.

It would be tempting, naturally, to reconcile the two points of
this conflict, to affirm both one's relationship to a community and
one's capacity to distance oneself from it. For the act of distancing
oneself is as indispensable to innovation as it is to a simple critique:
it too presupposes a certain *separation* from the reality that threatens
to submerge us. It is clear that any culture worthy of this name, any
work of breadth, is both *particular,* rooted in a determined space and
time, and *universal,* endowed with significance accessible to men outside the original community. In this way it becomes unique or *singular,* it develops "individuality," if it is true that the singular
reconciles the particular and the universal.

This synthesis is thus quite fruitful. It is necessary to take its
measure in order to better define the conceptions of the Culture-

Nature relationships that the various philosophies of ecology still embody today. Yet we cannot forget that of the three points that compose it—the particular, the universal, and the singular—it is the second, that of detachment, of liberty conceived as transcendence, that is peculiarly human. If we did not have the ability to detach ourselves from the traditional culture that is imposed upon us like a second nature, we would continue, like all animals, *to be governed by natural codes.* If we could not put this culture in perspective and adopt a critical viewpoint, which alone allows us to change it and inscribe it in history, our culture of origin would be akin to animal habits, and human societies would be as devoid of history as those of ants or termites. Tradition, reduced to the pure and simple transmission of the past, would be merely the *instinct* peculiar to the human species, just as programmatic as it is in the other animal species. It is on this precise point of the debate, it seems to me, that the Enlightenment marks a decisive advantage over romanticism.

It is important to appraise here what such a line of thought implies with respect to philosophical anthropology. It has often been cited as the origin of ethnocentrism, even the intellectual foundation of colonialism. It is easy to see why: from the perspective that Rousseau, and thus all revolutionary ideology, adopts, it is difficult to distinguish "primitive man" from the animal as clearly as one would hope. The continuation of the text confirms this fear: "Savage man, by nature committed to instinct alone . . . will therefore begin with purely animal functions." There seems to be some doubt as to the true identity of the "primitive."

Let us take this one step further. If contemporary ethnology has taught us one thing, it is that so-called primitive societies are, or at least see themselves, as societies without history. Better still, this singular orientation toward fidelity to the past comes to them, to repeat Pierre Clastres's formula, from the fact that they are above all societies "against the State," that is to say, essentially, against innovation. Governed by tradition, they attribute customs and laws to a religious origin, which says everything[2]—that it is nonhuman

---

[2] Marcel Gauchet developed the consequences of this Clastrian intuition in his wonderful book, *Le désenchantement du monde* (Paris: Gallimard, 1985).

and situated in a mythical time of which the elders, or even the ancestors, are heirs and guardians. Modern natural law, the philosophy of the Enlightenment and the French Revolution, culminate on the contrary in the *secular,* humanist idea, according to which law must be based in the will of individuals, which is to say, in political humanity constituted into a people. Diametrically opposed to the world of tradition, revolutionary action thus organizes itself around the notion of the future, since it generally presupposes that one is seeking to transform reality in the name of a plan or ideal that necessarily belongs to the future. Conversely, the role of the Indian territories described by Clastres is not to bring about change but to maintain tradition through fidelity to age-old principles, the betrayal of which would cost a man his life.

The eighteenth-century philosophers did not have the advantage of following the teachings of Lévi-Strauss. The idea that cultures could be "others," that there is no single yardstick of human history by which to measure other cultures, was foreign to them. The existence of societies living resolutely outside of European civilization, even resisting its encroachment, could hardly be perceived as anything but negative. We are better equipped to understand the breadth of the problem: If man is characterized by his history, what status could one attribute to these "primitive" societies which by every indication have not entered the world of history, thus the specifically human world of "perfectibility" and of separation from nature? Must we not ultimately accept their classification on the side of the animal kingdom and compare them, because they are without history, to the hives and termitaries which are "at the end of a thousand years what they were the first year of that thousand"?

Decidedly, this distinction between humanity and the animal kingdom seems to carry horrifying consequences in its wake, and this is what those who scorn modern humanism have exploited. A more in-depth examination should convince us that, here too, things are not as simple as they seem. The birth of the democratic universe must not be considered solely with respect to what followed it (Europecentrism, one might say), but also, and even particularly, with respect to what preceded it (the hierarchical, closed, and finalized world of the Ancients).

From this perspective, the question at hand can essentially be answered in two ways. The first borrows from the antique world the idea of hierarchy. There is, on the scale of beings, a high and a low as well as a continuity between all degrees. We go from God to man, from man to animals, from animals to vegetables, then to minerals. We accept that within each class there exists another hierarchy (from the wise man to the fool, from higher mammals to the earthworm, and so on). Additionally, in keeping with this continuity, we suppose the existence of intermediary beings: hence the eighteenth-century philosopher's passion for crystals, about which numerous treaties were written. For these crystals seemed to be the missing link between organized beings and minerals, just as carnivorous plants establish a link between animals and vegetables. It is within this framework that we must place the emergence of the problematics of the *Untermensch,* so marvelously analyzed by Alexis Philonenko:[3] the reemergence on pseudoscientific bases of an archaic form of thought on hierarchy. The "primitive"—*sub*man or "*super*monkey," according to Buffon's extraordinary formula—would be defined as one who, *stricto sensu,* situated *beneath* humanity but *above* animals, would fill the void between the two kingdoms.

Yet it is impossible to avoid racism and its political consequences if one subscribes to the belief that primitive man cannot attain authentic humanity due to his essence or nature: inferiority is inscribed in his makeup, in his definition, so that "civilization" will remain forever foreign to him. Transplant an African to Europe and he will remain an African for several generations at least, requiring, again according to Buffon, "one hundred and fifty to two hundred years to whiten his skin"!

But this was not the *Aufklärers's* response. In any case, not that of the greatest among them, Kant, who sketches what is no doubt one of the first nonracist explanations of the necessities that may have constrained certain peoples to dwell on the margins of civilization and thus, so it *seems,* of humanity. The fact that his explanation is mythical, and as such without value from a descriptive or scientific

---

[3] For example, in Alexis Philonenko, *La théorie kantienne de l'histoire* (Paris: Vrin, 1986).

point of view, makes it all the more interesting from a philosophical one. Combining history and geography, it begins with the premise that it was through war that the surface of the earth was inhabited. How else can one account for the fact that men resigned themselves to living in the hostile lands of the far North? What can we draw from this? First, that it was the people on the peripheries, to the North and to the South, who avoided the influence of the civilization developed, in the center, by Europe. Next, that in these two regions, *it was nature, not race, that constrained men to a marginal status.* In the North, because nature is so hostile that it is impossible to construct anything lasting. The Greenlander has little choice: necessity obliges him to live in houses of ice and eat the raw flesh of the animals he hunts. The culture, both of the earth and of the spirit, encounters such deplorable natural conditions that it cannot freely expand. Though the situation appears more favorable in the South, it is, in reality, worse: in the Caribbean, man lives amid such generous natural conditions that he has no need to embark on the foreign and uprooting adventure of culture. The climate allows him to go naked; the fruits of the sea and the earth spare him the work of animal breeding or agriculture.

Anyone who considers this myth with the Third World sensitive view of a twentieth-century intellectual will no doubt note that what is lacking is an awareness that Caribbean or Greenlander cultures, while "other" than our own, are nonetheless equally legitimate. But were we to interpret it in light of the break it makes with the ancient world, we would note on the contrary how it endeavors to invent a plausible explanation for the difference between primitive and modern man within a democratic, nonracist framework of thought. For the main requirement is posited: *that this difference is not inscribed in a definition, in a racial essence.* We are forced to agree with Musil that a "cannibal taken from the cradle to a European setting will no doubt become a good European and that the delicate Rainer Maria Rilke would have become a good cannibal had destiny, to our great loss, cast him at a tender age among the sailors of the South Seas."

It is now possible to measure with greater precision the way in which the distinction established by Rousseau plays a founding role in the birth of modern culture and ethics, how in particular it enables us to clearly and systematically reconstruct two essential poles: "good will," understood as the ability to act in a disinterested manner and the valorization of universal goals, versus the valorization of goals that are valid only for oneself. The notion of acting in a disinterested manner is absolutely foreign to the ethics of the Ancients. While virtue, as Aristotle believes, is the realization, for each being, of that which constitutes his *telos,* simultaneously his essence and his end, its actualization coincides far more with happiness than not. Yet once humanity defines itself by perfectibility, by the capacity to break away from natural or historical determinations, the bases of moral teleology collapse. Man (and then woman . . .) is devoid of a specific function. It is impossible to read any type of destination into his makeup, to identify in his nature the sign of any sort of plan. A woman, it is true, can bear children: that, to speak Sartre's language, is her *situation.* But to declare that she "is made to bear children" is to deny her liberty, to transform her situation into a *natural determination,* in short, to "animalize" her. And if the human being is "nothingness," if he possesses no "nature" in which some sort of "mission" can be discerned, virtuous activity can no longer be considered in terms of finality. Virtue is no longer the actualization of a felicitous nature. One can no longer speak, as did Aristotle, of the "virtue" of a horse or of an eye to designate its natural excellence. From then on the term is reserved for humans, for that which manifests a distancing from naturality, *whether we speak, in the ethical sphere, of renouncing personal interests in favor of the general good, or, in the cultural sphere, of separating oneself from the seemingly natural facts of one's national, linguistic, or other traditions.*

The most fundamental ethical requirement among Moderns, that of altruism, is in its very principle antinatural, since it requires a form of disinterestedness. It presupposes "good will" and is inevitably expressed in the form of an imperative. But the reference to universality, which is incomprehensible outside of the framework of this new philosophical anthropology, also becomes necessary. For the separation from historical-natural codes, through which man manifests his difference from animals, is still a refusal to allow oneself to be limited to any particularity. It is because he is capable of taking his distances not only from the cycle of his biological life but also from his particular language, nation, and culture that man can enter into communication with others. His capacity for universality is a direct function of this distancing.

The connection between the Judaic tradition and that of critical philosophy has occasionally been the subject of speculation. Horkheimer, for instance, has written beautifully on this theme. And since the death of Hegel, Judaism has in fact often expressed itself in the categories of Kantian thought when it needed to take the form of a philosophy. To denounce the Hegelian association of the Jewish conscience with the image of the "unhappy conscience," ever separated from the absolute and subjected to the exteriority of an imperious law, the Kantian critique of the pretensions of metaphysics was mobilized by every school from Marburg to Frankfurt. I believe that there is a profound reason for this: it is that for critical philosophy, as for Judaism, man is the antinatural being and, as such, the being who can live according to law. This, it seems to me, is what Lévinas perceived in these lines from *Difficile liberté:* "The Jewish man discovers man before discovering the countryside and the cities. He is at home in a society, before being so in a house. He understands the world based on those around him rather than the whole of being based on the earth . . . This freedom is not unhealthy, not twisted, not anguished. It places secondary importance on the values of rootedness and institutes other forms of fidelity and responsibility."

Still, the human condition, as distinguished here from the animal realm, continues to pose the problem of reification. For one is hard pressed to see how man could determine himself without becoming "someone," and without also becoming "something." One is at constant risk of identifying with a character, a personage, of assuming a social, familial, or sexual role, in short, of assuming an essential nature. This identity is even inherent to maturity, which it would be vain, but also absurd, to reject indefinitely. One can try to be ironic, manifest a *detachment* from all experience in order to reassert the power of absolute freedom, rejecting the alienation of a particular identity. It is doubtful that such abstraction would be satisfying, even supposing that it were possible. In any event, the ineluctable onset of old age would gradually reestablish the laws of nature over those of freedom. Whereas systems of traditional thought, entirely governed by references to the past, managed to confer significance on the fact of aging, we Moderns remain singularly at an impasse. If perfectibility alone is human, what sense can we make of the individual itinerary which, after following an ascending curve, inevitably approaches decline? One may argue—and this is the solution par excellence for the great philosophies of history—that the meaning of existence is situated in the species rather than in the individual, who merely makes a contribution to the edifice of a far greater whole. This solution is essentially a religious one, since it locates meaning beyond life. How can an individualist system of thought, more concerned with private existence than with the fate of the species, be content with it?

Hence the two crucial questions encountered by Promethean humanism when it admits, rightfully, that the faculty to separate oneself from the order of naturality is the sign of the properly human—the specific difference which is the source of all other significant or signifying particularities.

The first is that of the relationship to the natural universe. Is there not, from this perspective, radical antinomy between respect for nature and concern with culture? As if the choice of what is human could only be at odds with nature, and the choice of what is natural at odds with men, who, as bearers of excess and destruction, are identified with total evil. This would mean the victory of the hatred of men, in both senses of the expression; Rousseau's noble efforts would lead directly to the Cartesian plan for the devastating domination of the earth. In this case humanism *in all its forms* would have to be deconstructed and surpassed in order to create the possibility for taking ecological concerns into account.

But this first question depends on the fate of the second, that of the relationship to culture. Let us admit for a moment that freedom, understood as the ability to break away from the animal in us (for example, from the "natural" tendencies toward egoism) is, as Rousseau believes, the cultural faculty par excellence, without which culture would not be possible but only customs and "ways of life," like those that govern the animal kingdom. What concrete form will this freedom take if not the destruction—the perpetual uprooting— it seems to invite? To be "authentic," faithful to his essence (which is to have no essence), mustn't the human being *destroy* all content that would risk determining him? Is he not forced to affirm his freedom by dissolving any individual determination, by permanently rejecting all past tradition as well as any present incarnation? The humanist is cosmopolitan by vocation. But can he consider that which is local, or national, as something other than total evil? On a political as well as aesthetic level the dry, empty abstraction into which avantgardism has fallen in this fin de siècle represents the perfect illustration of the aporias of absolute freedom. The attempt to create a culture of separation culminates in the impossibility of *experience,* in the negativity of naked abstraction, in short, in the very opposite of what we have a right to expect from culture . . .

How can tradition, freedom, a concern for nature and the culture of humanism be articulated? Clearly, we cannot evade this line of inquiry, and will return to this question at the end of this essay. Beneath the seemingly banal quest for the features peculiar to the animal kingdom and to humanity, our position with respect to modernity is at stake. Because the latter is perceived by radical ecologists as tainted with anthropocentrism, they propose, with an eye to reinstating nature, to create new legal subjects. Beginning, of course, with animals. Can secular humanism rise to the challenge?

# "Animal Liberation," or
# The Rights of Creatures

Let us begin with a look at the facts; in France they are still perceived too confusedly for us to brush them aside. One might say that the question of the legal status and the protection of animals is constantly gaining in pitch. This is not entirely false, if by this one means that the fate of whales, ring doves, or baby seals occasionally mobilizes a segment of the public, which increases in receptivity as the number of domestic animals grows. And this number continues to grow, to the point of reaching unprecedented heights today:[1] in France, there are thirty-five million pets, including ten million dogs and 7.5 million cats, whose owners devote a budget of thirty billion francs every year to their various needs. This figure may be illuminating: in 1975 there was *one* veterinary clinic open twenty-four hours a day in the Parisian area. Now there are forty! Recently, and not surprisingly, we have seen the advent of animal cemeteries as well as scanner clinics, "dog-sitting" centers, kinesitherapy, balneotherapy, and psychotherapy for animals! Quite understandably our politicians are hard-pressed to avoid the subject.[2]

But this new democratic passion has its drawbacks. One of them is the number of dog bites, which give rise every year to 500,000 complaints, representing between .5% and 1% of medical

[1] For an original and interesting interpretation of this progression, see Paul Yonnet, *Jeux, modes et masses* (Paris: Gallimard, 1987).

[2] It is in the nature of demagogy, which in 1988 reached staggering proportions in the Mitterand-Chirac debate, details of which must be faithfully reported:

CHIRAC: As the mayor of Paris and as a man, I was angry with you at the time. In 1984, you more than doubled the tax on dog and cat food.

emergencies. Half of the victims (mostly children between the ages of two and four) remain scarred, and 60,000 are hospitalized, while the owner risks . . . a mere forty-to-eighty franc fine! This serves as fuel for the now common polemic between zoophiles and zoophobes.

But again, all this remains quite hazy and confused in France, streaked with strong emotions and loving feelings. No one really expects Brigitte Bardot to develop a coherent doctrine on the rights (and responsibilities?) of animals. On a philosophical and political level, we have nothing that can compare with the "animal liberation movement," which not only represents thousands of people in the Anglo-Saxon world but has been sanctioned by the university. In the United States, Canada, or Germany there is no keeping count anymore of the academic colloquia devoted to the metaphysical and legal status of animals. In the meantime, the problematics are not always limited to theory—far from it (though the intellectuals of the movement preach nonviolence). The Association of American Medical Colleges has logged more than four thousand cases of threats on the part of liberation militants, and a "raid" on the campus of the University of California in 1987 caused some three-and-a-half million dollars worth of damages to a veterinary center. More surprisingly still, the number of vegetarians in England, only .2% of the population in 1945, rose to 2% in 1980 and to 7% in 1991. Three quarters of them declare having stopped eating meat out of respect for animals. A Gallup Poll conducted for the *Daily Telegraph* reveals that 70% of sixteen- to twenty-four-year-olds claim to be in favor of a total ban, or at least very strict limitations, on animal experimentation.[3]

It is true that Anglo-Saxon thought has significantly laid the groundwork here, though from Plutarch and Porphyry to Schopenhauer, by way of Montaigne, Maupertuis, or Condillac, there has

---

MITTERAND (remembering a point scored by Giscard against him in another debate): You don't have a monopoly on affection for dogs and cats. I love them too!

See also the report by Eric Conan devoted to this subject in the 19–25 January 1990 issue of *L'Express.*

[3] See "Man's Mirror," *The Economist,* 16 November 1991.

certainly been no lack of "continental" philosophers to have argued in favor of greater respect for those termed by Michelet our "inferior brothers." One could even extend the list considerably, turning to the East, evoking the sacred animals of India . . . [4] But for the most part, aside from the cases of a few rebels, it has only been within a very specific philosophical tradition, that of utilitarianism, that such an appeal has taken the form of a call for animal *rights* and not simply for additional responsibility for men. If we wish to clarify the terms of this appeal, it is necessary to recall its absolute antithesis: Cartesianism, and its theory of animal machines, which elicited lively polemics even in France.

## The Animal and the Automaton

In Cartesianism, *horresco referens,* contemporary zoophilia finds its absolute foil: the perfect model of anthropocentrism which accords *all* rights to men and *none* to nature, including animals. Why such disdain, one wonders? Because the subject, the *cogito,* cannot be the sole and unique pole of meaning without nature being ipso facto divested of all moral value. Cartesian physics took to the task of eradicating the notion that the universe is a "great living being," of doing away with the animism or "hylozoism" that still dominated scholastic thought. The principles of alchemy had to be pulled up by the roots. Not only is nature not "animated" but it contains no occult power, no invisible reality that could explain the metamorphoses of lead into gold or man into a werewolf! The material world is without soul, without life, without even a force; it is entirely reduced to the dimensions of "extension" and motion. Thus no mysteries are inaccessible to human knowledge in this simple mechanism of objects that is the universe. And animals are no exceptions to the rule.

The proof is that they don't speak, even those animals like the magpie or the parakeet that have the ability and organs to do so. Their words, when they are spoken out of *mimesis,* are not a language but the product of machinery with neither soul nor meaning. And

[4] See L. Ferry and Cl. Germé, *Des animaux et des hommes: Une anthologie* (Paris: Hachette, forthcoming).

another indication, a more certain, though paradoxical one, is that the animal, *like all well-built machines,* "functions" better than man: "I know," writes Descartes, "that animals do many things better than we do, but that doesn't surprise me, *for even this serves to prove that they act naturally and automatically,* like a clock that tells time better than our own senses. Thus when the swallows arrive in the spring, they are no doubt acting as clocks." Buffon takes up this idea in his *Histoire naturelle des Orangs-Outangs* [Natural History of Orang-Utangs]: "The most marked characteristic of reason is doubt . . . but movements and actions that indicate only decision and certainty prove at the same time mechanicalness and stupidity."

What resulted, inevitably, was the following mode of thought: animals do not suffer, and their cries in the course of vivisection have no more meaning than the ticking of a clock. Such was the price for putting an end to the "vitalist" illusions of medieval Aristotelianism. This thesis also allowed many theological difficulties to be avoided: If animals suffer, while to all appearances they are not subject to sin, which requires freedom of choice, how can one "save" God from being suspected of injustice?

Yet let us beware of caricature. Descartes was less of an extremist than his disciples. Animals are still creatures of God, and there is a difference between a machine fabricated by the Creator and a human artefact:[5] any mechanic worthy of this name "will consider the body a machine which, having been crafted by the hands of God, is incomparably better regulated and has within it more admirable mechanisms than any that can be invented by man."

The Cartesians would pay no heed to this notice, taken as they soon were by automaton fever. From Vaucanson's famous duck, which was supposed to represent, in the words of its creator, "the motions of

[5] François Dagognet wrote beautifully on this subject in his long and exceptional preface to an edition of the *Traité des animaux* by Condillac (Paris: Vrin, 1987). I can do no better than to follow his arguments here. Large excerpts from most of the texts mentioned by Dagognet can be found in our anthology, L. Ferry and Cl. Germé, *Des animaux et des hommes.* I should also mention, on the same theme, Krysztof Pomian's interesting article, "De l'animal comme être philosophique," which appeared in *Le Débat,* no. 27 (November 1983).

the internal organs intended for the functions of drinking, eating, and digestion," to the flute players that captivated salons, everything was tried, or almost. Including the talking machine, actually a miniature organ, which Father Mersenne no doubt dreamed of slipping one day into the belly of a doll: "I am now occupied," he wrote in great confidence, on 15 July, 1635, "in figuring out how to make organ pipes pronounce syllables. I have already found the vowels a, e, o, and u, but i is giving me trouble, and then I found the syllable ve and fe . . . I am attempting to have an organ made that one could carry anywhere in one's pocket . . ."[6] Next came the first jointed protheses, the first player pianos, further incarnations of the uncanny attempt to think oneself God and do away with the mystery of life . . .

A reaction had to come, all the more hyperbolic as the original scheme was mad (which, by the way, doesn't prevent it from feeding the fantasies of our most eminent biologists to this day). It was against this form of Cartesianism that Maupertuis, one of the first in France in this context, would explicitly evoke "the rights of animals" endowed with sensitivity and intelligence. But we should also mention Réaumur and his *Mémoires pour servir l'histoire des insectes,* and Condillac and his *Traité des animaux,* who together founded a tradition that would be taken up by the republican humanitarians— Larousse, Michelet, Schoelcher, Hugo, and many others still—in the nineteenth century. At first it was a matter of showing that the animal is not a machine, that it thinks and suffers. As such, it possesses rights, or at the very least creates duties for humankind.

As a reaction against those who reduced animals to machines, here they were raised almost to the level of man. In a seesaw type logic, each side knew it was impossible to suddenly stop at the midpoint. Réaumur would take his position to new heights. The bee became a mathematician far superior to those of Antiquity. Witness the cells of the hive: "The more one studies the construction of these cells, the more one admires it. One has to be as skillful in geometry as we have become since the discovery of the new methods to know the perfection of rules which the bees follow in their work."

[6] See the texts quoted by Dagognet, in Condillac, *Traité des animaux,* p. 25.

This anti-Cartesianism would join forces in the nineteenth century with republican anticlericalism, for reasons that are easily understood: from this perspective, it is Cartesian-Christian *spiritualism,* with its excessive separation of body and soul, that is responsible for the disdain in which animals are held. A materialist—or at least pantheist—humanism, on the other hand, would see only a difference of degree, not of nature, between man and animals and would, therefore, be more inclined to protect the latter. Hence the fearsome imprecations directed by Clemenceau, in the wake of Michelet, against "The atrocious rhetors of the company of Jesus" who "deliberately ignore the bonds of nature that unite us to our brothers below" because they "push to nonsensical extremes the logic of the theory of the soul issued by God, the unique privilege of the human species."[7] It is also this perspective that prompted the Left to unflinchingly vote in favor of the "Grammont law" in 1850 (despite the fact that it was presented by a conservative deputy), a law which, for the first time in France, prohibited the public mistreatment of domestic animals.

We should nonetheless note the limits of this anti-Cartesianism, for the fact remains that it is of *humanist, and thus, in a sense, anthropocentrist* inspiration. For this very reason, the call for the respect of animals would rarely go so far as to recognize *rights.* The truth is it wasn't until 1924 that a proper "declaration of animal rights" appeared in France. It should be noted, moreover, that it was the work of an eccentric, André Géraud, and would have practically no repercussions. As Larousse himself points out, in the otherwise very anti-Cartesian and zoophilic article he devotes to animals in his dictionary, the latter are not "objects of justice," which is to say that they cannot be legal subjects comparable to humans. The call for respect would not move beyond the idea that our duties are nonreciprocal. Indeed, the phrasing used by Michelet ("our inferior brothers") and by Clemenceau ("our brothers below") are indicative of the exact reach of this "philanthropic" humanitarianism. It is true that animals *in and*

[7] Quoted by Maurice Aguilhon in "Le sang des bêtes," *Romantisme,* no. 31 (1981). Here I am returning to the thesis defended by Aguilhon in this excellent article, which in my view requires only one further point to be complete: there also exists a Christian tradition that favors respect för animals, which anticlerical republicans were naturally disinclined to mention.

*of themselves,* because they are sensitive beings and not simple machines, must inspire a certain compassion in us. But the most serious consequence of the cruelty and bad treatment inflicted on them *is that man degrades himself and loses his humanity.* That is why the Grammont law does not surpass the framework of anthropocentrism which contemporary zoophiles consider so unspeakable: the interdiction of bad treatment effectively applies only to *domestic* animals, which is to say those *close to man* (the law doesn't protect wild animals), and only places checks on cruelty inflicted *in public,* which is to say, if one thinks about it, *on cruelty that can affront or corrupt man's sensibility.*

Maurice Agulhon has put his finger on what distinguishes nineteenth-century zoophilia from the variety encountered today (notably in the Anglo-Saxon world):

> Today, animal protection could almost pass for a branch of ecology . . . In the nineteenth century, when one spoke of animal protection . . . , one had in mind almost exclusively, in any case principally, domestic animals, threatened by the violence of their masters, and it was hoped that by curbing this minor violence, one would help to curb the major violence of humans against one another. The protection of animals sought to be pedagogical, and zoophilia the school of philanthropy. It was a problem of our relationship to humanity, and not of our relationship to nature.[8]

To a large extent, this would also be the state of affairs in England during the same period. When William Wilberforce and Thomas Fowell Buxton founded their society for the prevention of cruelty to animals (RSPCA) in 1824, they were known, like Victor Schoelcher in France, for their "progressive" opinions in favor of the abolition of slavery, and they took every occasion to establish a parallel between the two matters. Their argument is simultaneously humanitarian and philanthropic. The first protective law in England (1822) went no farther than the Grammont law: it too merely prohibited the mistreatment of *domestic* animals *in public.* Man remains omnipresent.

[8] Ibid., p. 81.

For fundamental reasons, only a very particular philosophical framework,[9] that of utilitarianism, would allow anthropocentrism, whether Cartesian or anti-Cartesian, to be surpassed, giving a consistent—which is not to say incontestable—doctrinal form to the idea that man is not the only subject of law but rather, more generally, all beings able to feel enjoyment and pain.

## Utilitarianism and the "Animal Liberation" Movement

A founding grandfather, Jeremy Benthan; a founding father, Henry Salt; and a worthy heir, Peter Singer, an Australian university professor whom many consider to be the current leader of the cause: this, very schematically, is the genealogy which the "animal liberation movement" ascribes itself.[10] But its heroes aside, the essentials are clearly expressed in the doctrine.

Let us begin by putting aside a misunderstanding: utilitarianism is not, as one commonly held opinion mistakenly asserts, the theorizing of widespread personal egoism. It presents itself, on the contrary, as a universalism or, rather, an altruism whose principle could be expressed as follows: an action is good when it tends to generate the greatest sum of happiness for the greatest possible number of persons affected by this action. It is bad when it tends otherwise. Clearly, the initial postulate is so dissimilar from narcissistic hedonism as to enter directly into conflict with it: *there are cases which may require individual sacrifice in the name of collective happiness,* the exact nature of this conflict constituting, in fact, one of the main problems of utilitarian theory.

With this in mind, it is easy to understand that, based on such a premise, one would wish to extend legal protection to all beings

[9] With the exception of the case of "deep ecology" romanticism, which is addressed in the second half of this book.

[10] The work of Regan, Clarke, Linzey, Meyer-Abbisch, among many others, must also be mentioned. The American and German literature on animal rights is of a surprising abundance and wealth. A recent bibliography took more than six hundred pages to catalog it. I am doing no more here than giving an overview of the principles of applied utilitarianism, without entering into the details of the sometimes divergent philosophical arguments and positions of those who come out in favor of animal rights.

susceptible to suffering. Here we must quote an unabridged version of a passage that comes up repeatedly in zoophile literature, in which Jeremy Bentham expresses the founding idea of the entire movement. One will recall that he is writing at the time when France had just liberated its black slaves, while in British territories they continued to be "treated like animals."

> The day *may* come when the rest of the animal creation
> may acquire those rights which never could have been
> withholden from them but by the hand of tyranny. The
> French have already discovered that the blackness of the
> skin is no reason why a human being should be abandoned
> without redress to the caprice of a tormentor. It may one
> day come to be recognized that the number of legs, the
> villosity of the skin, or the termination of the *os sacrum*
> are reasons equally insufficient for abandoning a sensitive
> being to the same fate. What else is it that should trace
> the insuperable line? Is it the faculty of reason, or perhaps
> the faculty of discourse? But a full-grown horse or dog is
> beyond comparison a more rational, as well as a more con-
> versable animal, than an infant of a day or a week or even a
> month old. But suppose they were otherwise, What would
> it avail? The question is not, Can they *reason?* nor Can they
> *talk?* but, Can they *suffer?*

The central argument is clear: the specific differences ordinarily invoked to place value on humans to the detriment of animals (reason, language, etc.) are irrelevant. Clearly, no greater rights are granted an intelligent man over an idiot, nor the loquacious over the mute. The only significant moral criteria should be the capacity to experience pleasure and pain. From the start, the argument *is inscribed in a democratic framework:* in the tradition of Tocqueville, it counts on the progress of "the equality of conditions," so that, after the blacks of Africa, animals in turn enter the sphere of rights. The idea that we accord no greater rights to a wise man than to an idiot is also at the heart of a democratic world vision: an aristocratic system, on the contrary, would tend in favor of *proportional* rights and status.

We can now distinguish three opposing philosophical positions relating to the rights of animals:

—The Cartesian position, according to which nature, including the animal kingdom, is entirely deprived of rights in favor of the human subject, the unique pole of meaning and value.

—The republican and humanist tradition as it is outlined in Rousseau or Kant but also, in part, in nineteenth-century France. Four philosophical themes are intertwined here: man is the only being who possesses rights; the ultimate goal of his moral and political activity is not primarily happiness but freedom; this freedom, and not the existence of interests to protect, is the basis of legal order; notwithstanding, the human being has certain responsibilities toward animals, in particular that of not inflicting *unnecessary* suffering upon them.

—Utilitarian thought in which man is not alone in possessing rights, but rather all beings able to experience pleasure and pain. *Here we surpass the supreme principle of anthropocentric humanism:* the ultimate goal of moral and political activity is the maximization of the sum of happiness in the world and not primarily freedom; the goal of the law is to protect interests, whoever the subject that possesses the interests may be; *all other things being equal* (we shall see what this reservation signifies), it is just as wrongful to bring suffering to an animal as to a human being.

The book by Henry Salt, *Animal Rights Considered in Relation to Social Progress* (1894), would only clarify Bentham's ideas by applying them to the themes which to this day are the key points of zoophile literature: recognition of the rights of wild animals; criticism of slaughtering, hunting, animal experimentation, and the fashion for leather, feathers, or fur, and so on. While one cannot say that Salt's originality consists in establishing a bridge between the progress of democracy and that of the love of animals, since the theme is clearly already present in Bentham, it is true that Salt brings new vitality when he sets out to establish a strict logical connection between the existence of the rights of man and the necessity of instituting those of animals. "Have the lower animals 'rights?' Undoubtedly—if men have"—from the very first line of his book, this is Salt's conviction.

The idea is worthy of reflection and has become an absolute

commonplace of contemporary zoophile thought. Very close, even in its language, to Tocquevillian prophesies about the evolution of democratic societies toward an ever increasing equalization of conditions, this idea is expressed in Salt's writings as unshakable dogma: after the emancipation of slaves, then of women, it will soon be the animals' turn, so true is it that "the mockery of one generation may become the reality of the next." This is a necessity, one might say, that is inscribed in the "direction of history." For "from the great Revolution of 1789 dates the period when the world-wide spirit of humanitarianism, which had hitherto been felt by but one man in a million . . . began to disclose itself, gradually and dimly at first, as an essential feature of democracy."

He reflects in often subtle fashion, in a manner still close to Tocqueville, on the distance that separates the worlds of ancient and modern thought, as well as on the overturning of received ideas which the advent of democracy continuously produces within the heart of modernity. Salt gives indication of this which would ultimately play a considerable role in posterity: one year after the publication by Mary Wollstonecraft of her *Vindication of the Rights of Women,* an anonymous pamphlet appeared entitled *A Vindication of the Rights of Brutes,* (1792). Written by a Platonist philosopher by the name of Thomas Taylor, this lampoon meant "to evince by demonstrative arguments the perfect equality of what is called the irrational species to the human." It is derisive, of course, because as we are soon given to understand, it is a matter of a demonstration by absurdity, the central argument of which could simply be put: "If women, then why not animals . . . "

Taylor didn't know how apt his words would prove. For the same theme would not only be adopted by utilitarian partisans of animal rights, but also by a number of militant anti-Platonist feminists. Is it not due to this catastrophic dualism of body and soul, of the intelligible and the sensible, that nature, women, and animals, all three on the wrong side of the divide, would be subject to domination by the male fraction of humanity? "Phallo-logo-centrism," say those in the know . . . The excellent Professor Stone, whom we already encountered in the role of advocate for trees, would express this in a phrase that summarizes a good number of books: "We have been making persons of children although they were not, in law, always so. And we have done the same, albeit imperfectly some would say, with prisoners, aliens, women (especially of the married variety), the insane, Blacks, fetuses, and Indians."[11]

The fact remains that, through this theme of equalization of conditions, one of the traits most typical of the democratic universe is also attributed to animals, and that is *individualism* and the associated right to *authenticity*. Salt is no doubt one of the first, if not the first, to explicitly formulate this requirement with respect to animals: "To live one's own life—to realize one's true self—is the highest moral purpose of man and animal alike; and that animals possess their due measure of this sense of individuality is scarcely open to doubt." It is no longer a matter of protecting "our inferior brothers" from the bad treatment continually inflicted upon them by humans, but of demanding for them the right to a good life, to a full blossoming of the self.

In his book entitled, significantly, *Animal Liberation* (naturally the title evokes Women's Liberation), Peter Singer would give this new vision of the animal kingdom its most complete and coherent treatment. One point should be mentioned: Singer's readers know that he expresses reservations, quoting Bentham in fact, about the idea of rights in general. He prefers the term *animal welfare* to that of *animal rights,* and one often finds Singer contrasted with authors such as Tom Regan, for instance, who are more apt to express themselves in legal terms. But we mustn't forget that Singer and Regan have published together and agree on the essentials. We also shouldn't forget the connection between Singer and Salt, who is constantly talking about the "rights of animals." The truth is, it is only a matter of word choice: Singer, like Bentham, simply refuses to enter into a semantic conflict on the notion of rights, which strikes him mainly as "convenient political shorthand" to designate a broader idea: that of justice or of moral respect due a being of any kind. There are no doubt differences between Regan and Singer (on the "right to kill," to which I will return). But what matters here is that Singer, along with Regan, considers the animal as *worthy of respect in*

[11] Christopher D. Stone, *Should Trees Have Standing?* (Los Altos: William Kaupmann, Inc., 1974), p. 4.

*and of itself.* He takes it to be a "moral subject," endowed with *intrinsic* dignity. This, of course, is key.

The main ideas of Singer's book can be collected under four chapter headings:

## THESIS I: INTERESTS AS THE FOUNDATION OF MORAL RESPECT AND CRITERIA FOR THE LEGAL SUBJECT

It is the capacity for experiencing pleasure or pain that determines a being's dignity and makes it, in a broad sense, a legal subject. This capacity is translated as the fact of "possessing interests"—by which we see that utilitarianism differs both from anthropocentrism (man is not the only one to fulfill this condition, and thus to be a subject of law) and from deep ecology, since its definition of the legal subject excludes rocks and trees.

> The capacity for suffering and enjoyment is *a prerequisite for having interests at all,* a condition that must be satisfied before we can speak of interests in a meaningful way. It would be nonsense to say that it was not in the interests of a stone to be kicked along the road by a schoolboy. A stone does not have interests because it cannot suffer . . . A mouse, for example, does have an interest in not being kicked along the road, because it will suffer if it is . . . So the limit of sentience (using the term as a convenient if not strictly accurate shorthand for the capacity to suffer and/or experience enjoyment) is the only defensible boundary of concern for the interests of others. To mark this boundary by some other characteristic like intelligence or rationality would be to mark it in an arbitrary manner.[12]

If law is, in a broad sense, the system by which interests are recognized and respected, rocks and trees are excluded from it. Note, on another front, the fundamental difference that separates utilitarianism from the humanism inherited from Rousseau and Kant: for the latter, *it is, on the contrary, the ability to separate oneself from interests (freedom)*

[12] Peter Singer, *Animal Liberation,* 2d ed. (New York: A New York Review Book, distributed by Random House, 1990 [1975]), pp. 7–8.

*that defines dignity and makes the human being alone a legal subject.* This humanism is thus the antithesis of Singer's position; the latter mistook its true nature from the start. He commits an obvious error—which is worth examining—when he believes he detects among his adversaries a concern solely with giving priority to reason, language, or intelligence. For it is clear that, in the eyes of Rousseau and of Kant, it is not these latter qualities that qualify man as a moral being but freedom or, on another level, "good will," which is to say the capacity to act in a nonegotistical or, literally, *disinterested* fashion.

## THESIS II: ANTISPECIESISM OR THE FORMAL EQUALITY OF ALL BEINGS WHO EXPERIENCE SUFFERING AND/OR ENJOYMENT

All things being equal, one interest equals another, given that "an interest is an interest, no matter whose interest it is." This principle establishes a *formal* absolute equality among all beings capable of experiencing enjoyment and pain. As Singer says in a formula that summarizes his "antispeciesism" and is the foundation for his tireless criticism of "human chauvinism": "All animals (including man) are equal."

Here we encounter the argument that made Salt's book so original: it is because we admit the validity of antiracist and antisexist arguments, because we are, from a rational point of view, forced to recognize the equal value of interests as such (the interests of blacks equal those of whites; those of women equal those of men) that we must take an extra step and accept the fundamental principle of antispeciesism:

> Speciesism—the word is not an attractive one, but I can
> think of no better term—is a prejudice or attitude of bias
> in favor of the interests of members of one's own species
> and against those of members of other species. It should
> be obvious that the fundamental objections to racism and
> sexism made by Thomas Jefferson and Sojourner Truth
> apply equally to speciesism . . . Racists violate the principle of equality by giving greater weight to the interests
> of members of their own race when there is a clash between

their interests and the interests of those of another race. Sexists violate the principle of equality by favoring the interests of their own sex. Similarly, speciesists allow the interests of their own species to override the greater interests of members of the other species. The pattern is identical in each case.[13]

The equality defended is thus none other than the formal equality found in the liberal declarations of the rights of man. Singer rightfully insists: these canonic texts consider that "equality is a moral idea, not an assertion of fact . . . *the principle of equality of human beings is not a description of an alleged actual equality among humans: it is a prescription of how we should treat human beings.*"[14] As in Bentham and Salt, but also in keeping with the entire tradition of political liberalism, it is a matter of recognizing that individuals are equally respectable *by law,* whatever the differences of intelligence or talent that distinguish them de facto. The fact that one may possess an IQ that is higher than the mean, be better spoken, better looking, taller, or more creative, confers no additional rights on the fortunate beneficiary of these talents. To parody one of Bentham's famous examples, the uncultured peasant would enjoy the same legal guarantees as Isaac Newton should they happen to find themselves on opposing sides of a lawsuit. Thus the fact that humans are (generally) more intelligent than animals confers no *intrinsic* superiority to the value of their interests.

We begin now to see how Singer's position is more coherent and stronger than that of the ordinary zoophile. It cannot be reduced to the status of a simple symptom of a sociological reality. The progress of democracy and that of respect for living beings may be connected, but it is also the case that the arguments invoked by Singer are essentially already present, at least in principle, in Bentham's utilitarianism—well before the love of animals become a mass passion. It should also be observed that, within the conceptions of law that are dominant in Anglo-Saxon countries (where the legal system is considered a system that protects interests), it is not at all easy to find convincing

[13] Ibid., pp. 6–9.
[14] Ibid., pp. 4–5.

arguments to oppose the bridge Singer establishes between the rights of man and those of animals. Indeed, from this perspective, how can one contest that "an interest is an interest, no matter whose interest it is"? How can one fault Singer for drawing the consequences of premises that seem, all in all, widely accepted as valid (although there are attempts to refute them, John Rawls's *A Theory of Justice* being the best example).

## THESIS III: DIFFERENCES BETWEEN ANIMALS AND MEN

If we wish to consider the antispeciesist thesis at its best, we should add that formal equality among all animals, whether "human or not," in no way implies a lack of differentiation between particular cases. In the name of utilitarianism we must accept that, if certain beings suffer more than others under certain conditions, they must be treated differently—the important thing is that this difference does not depend on one's belonging to a given species a priori but rather on the reality of the suffering.

To take an example dear to Singer: a human condemned to death will suffer more than an animal placed in the same situation, for he will anticipate his sentence. On the other hand, a wild animal placed in "prison" (a cage) will suffer more than a human being, for it is impossible to make him understand the ultimately provisional nature of his confinement, or that no harm will come to him. It is, therefore, necessary to adapt the principle of formal equality, so that what is legitimate for a man will not be so for an animal and vice versa:

> Precisely what our concern or consideration requires us to
> do may vary according to the characteristics of those affected
> by what we do: concern for the well being of children grow-
> ing up in America would require that we teach them to
> read; concern for the well being of pigs may require no more
> than that we leave them with other pigs in a place where
> there is adequate food and room to run freely. But the basic
> element—the taking into account of the interests of the

being, whatever those interests may be—must, according
to the principle of equality, be extended to all beings, black
or white, masculine or feminine, human or nonhuman.[15]

The same slap applied to a baby's behind or to that of a horse will
not inflict the same level of pain and will not, therefore, have the
same degree of illegitimacy—even though we might imagine the existence of an equivalent blow that would produce an identical effect
on the horse and on the baby . . .

## THESIS IV: THE END OF ANTHROPOCENTRISM

By virtue of the preceding, the principle of human privilege vanishes.
A prerogative could be accorded to it only *in certain cases* and for ra-
tional reasons (able to be calculated in terms of pleasure and pain).
Indeed:

> Whatever the criteria we choose, however, we will
> have to admit that they do not follow precisely the bound-
> ary of our own species. We may legitimately hold that
> there are some features of certain beings that make their
> lives more valuable than those of other beings; but there
> will surely be some nonhuman animals whose lives, by any
> standards, are more valuable than the lives of some hu-
> mans. A chimpanzee, dog, or pig, for instance, will have a
> higher degree of self-awareness and a greater capacity for
> meaningful relations with others than a severely retarded
> infant or someone in a state of advanced senility.[16]

Let us be fair to Singer: if he selects these examples, it is to show
that, whatever the criteria chosen—conscience, rationality, sociability, or the capacity for enjoyment and pain—we need to abandon
speciesism *because of the fundamental continuity between the animal species
and the human species, a continuity by virtue of which, in the most extreme
cases, which are more frequent than one would think, the animal turns out to*

---

[15] Ibid., p. 5.
[16] Ibid., p. 19.

*be superior to man, according to all possible pertinent criteria.* (I will nonetheless note that Singer *never* considers the criteria of freedom defined as the faculty to separate oneself from nature, to resist selfish interests and inclinations—a criteria that propels the entire tradition issued from Rousseau and Kant to distinguish animality from humanity.)

Hence the incessant examples proposed by Singer to test his theory: between the mentally defective infant, the senile old man, and the healthy pig, whom must one choose? Here, among many others, is a sampling of this type of problem:

> Adult chimpanzees, dogs, pigs, and members of many other species far surpass the brain-damaged infant in their ability to relate to others, act independently, be self-aware, and any other capacity that could reasonably be said to give value to life. With the most intensive care possible, some severely retarded infants can never achieve the intelligence level of a dog. Nor can we appeal to the concern of the infant's parents, since they themselves, in this imaginary example (and in some actual cases) do not want the infant kept alive. The only thing that distinguishes the infant from the animal, in the eyes of those who claim it has a "right to life," is that it is, biologically, a member of the species Homo sapiens, whereas the chimpanzees, dogs, and pigs are not. But to use *this* difference as the basis for granting a right to life to the infant and not to the other animals is, of course, pure speciesism. It is exactly the kind of arbitrary difference that the most crude and overt kind of racist uses in attempting to justify racial discrimination.[17]

A partisan of euthanasia (in the name, of course, of the same philosophical options: Why continue to live if total suffering is greater than total enjoyment?), Singer is to this day banned from speaking in Germany, where his words have been interpreted as

[17] Ibid., p. 18. For other examples, see p. 19 and Singers's many works of applied ethics.

a veritable call for the murder of the mentally ill. The repressed guilt, the memory of the infamous "past that won't pass," are apparently at the origin of this hostile attitude. Intellectual honesty nonetheless forces us to note that Singer's theses are only, here again, a logical and coherent consequence of the fundamental premises of utilitarianism. I therefore suggest to the German university professors who refused to allow Singer to speak (or who, at the very least, gave in to pressures that tended in this direction) to immediately demand the burning of the books by Bentham and Stuart Mill . . .

To these four fundamental theses, let us add two further reflections which must be taken into account in the discussion: one regarding the "right to kill," the other regarding the pressing issue of the animal liberation movement.

Contrary to an opinion one might consider self-evident, utilitarianism, consistent with its premises, is not always hostile to all forms of killing a priori, whether in the case of humans (as we just saw) or of animals. It is on this point, notably, that Singer diverges from other zoophile theoreticians with whom he is otherwise very close in view, such as Tom Regan. A professor at the University of North Carolina, author of a famous work entitled *The Case for Animal Rights* (1983), Regan defends the notion that the animal is "the subject of its own life." As such, it *possesses* particular qualities such as memory, beliefs, preferences, emotions . . . and among these, the right to live, that is to say, concretely, the right not to be deprived by others of the pleasure of its own future—which is why, among humans, Regan notes, we have greater regrets about the death of a young person than of an old person, since the former is being deprived of more than the latter.

As for Singer, he is content to defend strict utilitarianism: in certain situations, death by suicide or euthanasia may be considered preferable to life. The decision belongs with the subject. Similarly, in certain cases one must resolve to sacrifice animals so that humans can live—the classic example being that of the "native Americans" who had to hunt buffalo to feed themselves in winter. It is all a matter of prudence, of practical choices made as a function of contingent casuistry. The only principle that must be asserted is the following: "As long as we remember that we should give the same

respect to the lives of animals as we give to the lives of those humans at a similar mental level, we shall not go far wrong."[18]

The second remark is that the movement's pressing issues are
unrelated to media-based sentimentality, but only to the calculable
and calculated tally of the quantity of suffering occasioned by our
behavior toward animals. In other words, neither the whales lost
in the ice, nor abandoned dogs, nor even spotted owls or baby seals
are emergencies, strictly speaking. But what is urgent for Singer are
(1) the tens of millions of laboratory animals sacrificed each year,
and (2) the billions of animals raised as livestock for food. Singer's
constant concern with coherence is maintained when it comes time
to move into action.

## Elements for a Critique of "Animal Liberation"

The main problem with Singer's theses—if we put aside their potential for ridicule due to the absurd or shocking nature of certain of
their deducible consequences—lies in the fact that he accepts as the
ultimate horizon of rationality the calculating logic of interests
alone. It is unfortunate, under these conditions, that he never correctly discusses the points of view of Rousseau and Kant, whom he
wrongly classifies as among those who consider man's superiority over
animals as strictly dependent on reason. This is to overlook the entire
difference separating their critical thought from that, for example, of
Aristotle. For it is freedom (or, to use Kant's terminology, "practical"
reason) and not intelligence (theoretical reason) that constitutes the
specific difference being sought. It is possible to contest the notion,
as Singer does, that there is discontinuity between men and animals
in their capacity to experience suffering. As far as this goes, everything is no doubt a matter of degree, not of qualitative leaps. But this
assertion does not respond to the position that places the foundation
of man's dignity elsewhere, notably in freedom and not in natural
sensation.

One may perhaps respond, as does Bentham, that only suffering
matters. But a second difficulty arises: for it is necessary to show in
what way the suffering of animals is worthy of respect as such—a

[18] Ibid., p. 21.

demonstration which utilitarianism spares itself the trouble of successfully completing, convinced as it is that the existence of interests immediately and obviously establishes, at least in principle, the protective right of these interests. Hence it endeavors to prove that animals have interests (since they suffer), thus, broadly speaking, rights (since they have interests), without paying attention to the fact that it is not the content of the argument itself (animals have interests), but its fundamental premise (interests are the basis of law) that a disciple of Rousseau would contest. In one of the texts he devotes to the question, Singer believes he has found a solution with the following argument:

> Other things being equal, it cannot be in my interests to suffer. If I am suffering, I must be in a state that, insofar as its *intrinsic* properties are concerned, I would rather not be in . . . Conversely, to be happy is to be in a state that, other things being equal, one would choose in preference to other states. There may, of course, be other things that we value, or disvalue, besides happiness and suffering. The point is that once we understand this method of ethical reasoning, the significance of suffering and happiness is indisputable.[19]

The reasoning is sound but in no way shows the *ethical* significance of suffering, only its importance for us, which is not even in question. In other words, many things may be important to me but not necessarily for moral reasons. No one will deny that love is the source of both the greatest joys and the greatest sorrows. And yet, it is unclear whether morality, or for that matter law, have a place in it, whether their rules regulate its fate . . . Let me make myself clear: I am not saying that suffering cannot have ethical significance in certain cases, that we should not, for example, strive as much as possible for the happiness of others, or at least not to cause them unjustified pain. But I am simply saying that this notion is not self-evident, and that it is necessary to further develop the argument, to better define the ethical weight of enjoyment and pain so that the call for a certain respect

[19] Peter Singer, "The Significance of Animal Suffering," in *Behavioral and Brain Sciences* 13 (1990), p. 11 esp.

for animals, less dependent on a particular (utilitarian) doctrine, can achieve better footing on a philosophical level.

For want of such an analysis, we are obliged to fall back on a simple fact of *nature*—hence the animal liberation movement's kinship with radical ecology. Though different from deep ecology, the utilitarian position is nonetheless infused with a form of antihumanism. Unlike deep ecology, though, it aims to show that animal rights are in direct line with human rights, with the rational and democratic refutation of racism and sexism. *But culture, understood as an outgrowth of freedom, itself defined as separation from nature, is never taken into account as such. For if everything is calculable, according to the logic of utilitarianism, it is precisely because everything is natural.* This is why there can be no *discontinuity* between nature and culture, between animal-kind and humankind.

Which brings us to a final question, this time forcing us to address the delicate problem of the internal coherence of utilitarianism: if in effect we consider that everything is calculable, that all one need do is account for interests and that only interests are worthy of consideration, on what grounds can one appeal for *sacrifice,* which a moral position more or less always presupposes? This is in fact a classic line of questioning, one that even preoccupied the founding fathers of the doctrine—beginning with Henry Sidgwick[20]—yet which Singer seems to entirely ignore. What utilitarianism considered to be virtuous was not selfish action, which takes into account not only personal interest but action that considers the global sum of suffering and pain in the world. Nonetheless, how can one move from one to the other if everything is but calculation? Why would I accept to no longer eat fine foie gras if I am indifferent to the suffering of the overstuffed geese, if I feel no *sympathy* for them? In short, if I do not have the capacity to act in an *antinatural* manner—the freedom of which Rousseau spoke—if I do not have to separate myself from selfish interests, to rise to more global considerations, why would I obey utilitarian principles? What is more, Is it not this very faculty of freedom that alone allows me to posit moral values *and to distinguish*

[20]On this point, see the interesting article by Lukas Sosoe, which appeared in the *Cahiers de Philosophie Politique et Juridique de Caen,* no. 18 (1990).

*them from simple interests which, as long as they are not mine, may rightfully
leave me indifferent?* Is this not where the qualitative, and not simply
quantitative, difference between men and animals lies? We have seen
men sacrifice their lives to protect whales; it must be said that the
reverse is far less common.[21] And supposing this were not a simple
coincidence but rather an essential difference, a difference between
beings of nature and beings not only of reason but of freedom, which
is to say of antinature?

Singer emphasizes, perhaps legitimately, that the fact of possessing language, mathematical reasoning, or more developed sociability
or affectivity are not sufficient grounds for qualifying a creature as a
moral being. On the other hand, it is difficult to see how, in the absence of freedom, any normative ethic could be put into effect—but
only an *ethnology,* a simple description of mores and customs, incapable by definition of being in any way prescriptive, of creating the
imperative force of respect which, precisely because it is imperative,
always contains within it an element of opposition to nature.

The criterion of freedom turns out to possess an entirely different
status from the other criteria invoked by Singer (reason, language, sociability, and so on). For if we follow Rousseau, who is entirely anti-
Cartesian (animals have sensibility) and anti-Aristotelian (the specific
difference between men and animals is not reason, for animals clearly
possess intelligence), the continuity between living beings on which
Singer's argument essentially rests ceases to be as incontestable as he
imagines. No doubt one could demonstrate that there is a certain continuity in suffering, intelligence, or even in language, but when it
comes to freedom, men and animals seem to be separated by an abyss
which bears that same name—history—whether we look at the individual (education) or the species (politics). *Until the existence of proof to
the contrary, animals have no culture, but only customs and modes of life, and
the surest sign of this absence is that they transmit no new legacy from generation to generation.* That is, unless we consider, as does sociobiology, that
human culture is also merely an outgrowth of a certain nature (but
under these conditions, why does it evolve? Why isn't there just one

---

[21] Which does not exlude the possibility of animal devotion tied to
*affectivity.*

culture for the species, as is the case for bees or ants?), one is forced to account for this specific difference, this radical discontinuity.

How can one respond to the question constantly put forth by Singer: In the name of what rational, or even only reasonable, criterion can one claim that humans must be respected more than animals in all cases? Why sacrifice a healthy chimpanzee over a human reduced to a vegetable state? If one were to adopt the criteria that says there is continuity between men and animals, Singer might be right to consider as "speciesist" the priority accorded human vegetables. If on the other hand we adopt the criteria of freedom, it is not unreasonable to admit that we must respect humankind, even in those who no longer manifest anything but its residual signs. Thus we continue to treat an important man with respect for what he was in the past, even when the impairments of age have long robbed him of the qualities that may have made him a genius artist, intellectual, or politician. For the same reasons, we should put the protection of cultural works above that of the natural modes of animal life even though, fortunately, the two are not mutually exclusive. An ethical preference for the reign of antinature over that of nature does not prevent us from considering and, if possible, doing right by the enigmatic characteristics of the animal kingdom.

*three*

## Neither Man nor Stone:
## The Enigmatic Being

The photograph shows a bull being followed by a disturbing crowd.
It is immediately apparent that he is going to die. But as the scene
evokes a lynching ritual, we sense that the road to this final liberation
will be long and painful. This strange drama unfolds in our time, in
Spain, in a little village called Coria. It is a game, of course, the rules
of which are extremely simple: at four in the morning, a bull is set
loose on the streets, where he is riddled with darts, first in the eyes
and most sensitive parts. Four hours later, the animal is beaten until
he dies of his wounds. In the picture, he looks like a pin cushion. The
little white dots are so close together he almost looks snow-covered.
Two men are pointing at him, smiling. One of them holds a ban-
derilla which he will plant, several seconds later (as appears in another photo), in the animal's anus. The entire population participates
in the festival . . .

This game has nothing, or almost nothing, to do with bullfighting. No particular talent is required to participate, and no one would
dream of comparing it to an art. It is simply an enactment, solely for
entertainment, of the reality of animal suffering. And people find this
suffering captivating. The proof being that this type of entertainment, in which crowds gather to see the intensity of the pain preceding the killing, has its equivalent in every, or almost every, country
and at every period. In France, during the Restoration, people sought
entertainment in open-air cafés near the gates of Paris:[1] here a rooster
is stoned to death, there an archery match is held in which the target
is a live rat nailed to a wooden plank. In our day, in Australia, the

[1] See Aguilhon, p. 83.

rabbit population becomes a real scourge from time to time. But is it necessary to organize baseball games in which the animal, replacing the ball, literally explodes on impact with the bat, to the vast enjoyment of large, enthusiastic crowds?

People like to talk about the cruelty of Chinese markets, where chicks and kittens are "cleaned," placed on a spit and grilled alive, snakes are cut into slices while being kept alive for days on end to better preserve their flesh, heads of monkeys are drilled open in order that one may enjoy the warm brain as the animal continues weakly to struggle . . . Literally and figuratively, there is nothing prohibiting us today from continuing to torture "nonhuman beings," since they are merely unimportant heaps of matter. And if, for one reason or another, they are unlucky enough to be classified as "vermin," their "destruction" even becomes a legitimate and useful exercise. The final prohibitions fall away, and any means become valid to arrive at ends condoned by both the public authority and the ambiant Cartesianism. The animal caught in a trap, captured alive, will not end his days sweetly: alone against man, who possesses all rights over it, custom has it that it will pay dearly for belonging to the realm of those considered noxious to the masters of the earth.

Even in the case of stock farming, the mere fact that the animal is destined to be killed for consumption is almost always enough to ensure that it will have a bad time of it, *because as mere raw material, it is deprived of all dignity.* Thus the chickens plucked alive, the live frogs whose legs are removed, the rabbits whose eyes are pulled out for bleeding, the pigs who are beaten before being slaughtered ("it makes them taste better"), if possible slowly and painfully . . . Certain peasants are kind enough to kill them first, but they are less common than one thinks, and in any case they are not obliged to do so. I still have a cookbook, published in France between the wars, which specifies that for a certain recipe to be successful "the hare must be skinned alive." A strange requirement indeed . . .

We cannot help but wonder: Why so much hatred if animals are only things?

To answer this, we need a phenomenology of the enigmatic nature of animals and of the contradictory sentiments it evokes in us, particularly in a modern world in which the hierarchies between beings have grown hazy. Psychoanalysis has no doubt taught us a great deal about the nature of sadism as a disposition of the subject. It has taken less interest, on the other hand, in the *objects* upon which this sadism most readily alights, the first among them being the living body of the animal.

Here is why, I believe, its enigmatic nature so fascinates us.

We are, at least since Descartes, *authorized* to treat animals as simple *things* devoid of the slightest ethical significance. The Grammont law protected *domestic* animals against cruelty inflicted *in public*. But the legislation, which is still fundamentally Cartesian, doesn't even mention wild animals or private brutality. This is to suggest that they have no moral status and that in treating them "like animals" our "super-ego," or whatever stands in place of it, is safe. But what pleasure (or what terror), what unavowable private quiver would come over us if we were *really* only dealing with machines? Maupertuis already said it, speaking against Descartes in a famous letter: "If animals were pure machines, to kill them would be a morally indifferent but ridiculous act: like smashing a watch."

The remark is more profound than it seems. Animals are neither automata, nor plants with roots "in the belly," as the Cartesians had it. The truth is we know they suffer. And even if we cannot evaluate the exact nature of their pain, the degree of consciousness attached to their cries, the signs are sufficiently visible, the symptoms sufficiently transparent for there to be no doubt as to their similarity to those of humans.

Now, the spectacle of suffering cannot leave one *entirely* indifferent, whether it is a matter of a pig or a rabbit. For it is, according to a certain conception of life, the ultimate symbol of not belonging to the world of things: it is *finalized,* it induces reactions, such as flight, which is evidence of a *significance.* At the end of the eighteenth century, life was apt to be defined as "the faculty to act according to the representation of a goal"—which is why it was believed that plants, which cannot move "because they have their stomachs in the earth," were not living beings. This definition no longer has a place within the structure of contemporary sciences. It nonetheless continues to have meaning from the perspective of a phenomenology of the signs of freedom: finalized movement, or action if one prefers, remains for us the visible criteria of animal nature, what distinguishes it from unorganized matter, but also from the vegetable world—which is why the intermediaries, anemones or carnivorous plants, are still somewhat mysterious to us. And it is because of this capacity to act in a nonmechanical fashion, oriented by a goal, that the animal, *analogon* of a free being, appears to bear a certain relationship to us, whether we like it or not. Its suffering is evidence of this. *Thus considered in nonutilitarian fashion, this suffering furnishes a synthesis between the idea that one must respect animals in order not to debase man and the idea that animals possess intrinsic rights.* This is the significance of the *analogy*.

Thus it is not a matter of returning to and contesting the fundamental difference established by Rousseau between animal-kind and humankind. The animal remains a creature of nature. Yet how can we help but account for that part of it that is not merely a thing? How can we deny its enigmatic nature? In this respect, suffering is but one of many signs that allow us to see that the animal and the automaton cannot be included in the same classification of beings.

This is no doubt why animals can arouse both sadism and compassion in us—sentiments that appear to be the two faces of a same psychic disposition. The animal is the *dreamed* object. We should recall here what Freud teaches us about "waking dreams": they are the stories one tells oneself in order to satisfy, at least in an imaginary fashion, certain desires left unsatisfied in reality. Hence the banal yet embarrassing nature of these "castles in Spain" in which the hero, which is to say oneself, is overwhelmed by the success denied him in the real world. Money, power, love: he has everything, almost to excess. As the saying goes: "The neurotic builds castles in Spain, the psychotic lives in them, the psychiatrist collects the rent . . ."

In dreams, then, the plot corresponds to our desires. But when these desires are *forbidden,* because they are reputed to be immoral or inappropriate—and Lord knows this is common—one has to cheat: the storyline must be "elaborated upon," sufficiently deformed, encoded, for our moral standards to lose their bearings—yet clear enough, of course, for our desires to be fulfilled. Thus, dreams evade the vigilance of the superego to more calmly address the libido. This is what accounts for their strange obscurity.

Whether this theory is "true" or not is largely irrelevant here. It offers a model that allows us to understand the enigmatic nature of animals and, with it, the complexity of the sentiments of sadism or compassion it evokes: by asserting that animals are things, simple mechanisms, we evade the prohibitions that weigh upon possible sadistic impulses. By acting in this fashion, we do more than legitimize the impulses: we eliminate them, since there is no such thing as sadism toward inanimate objects. But since we are more or less covertly aware that in truth animals are not entirely things, that as luck would have it they suffer, the tortures we inflict remain interesting.

If we raise our sights from the phenomenology of feelings to philosophy, should humanism itself not be incriminated? Are we not justified in discerning in it a formidable ideological enterprise aimed at legitimizing, by way of "rationalization," the colonization of "brute" nature and more, of the living in all its "prehuman" forms? Is it not humanism that dictates that compassion toward animals must be ridiculed at any price and qualified as infantile "sentimentality?" The question cannot be rejected out of hand but warrants closer examination.

## Is Humanism Zoophobic?

The hypothesis of a secret complicity between humanism, which is necessarily anthropocentric, and the exploitation of nature does not seem particularly original. Within the context of contemporary thought, it is a standard argument among critics of modernity. There are neo-Marxist versions of it in Adorno's and Horkheimer's *Dialectic of Enlightenment,* and neo-Heideggerian versions are even easier to find. It is this second path that Elisabeth de Fontenay chose to follow in an exemplary article entitled "La bête est sans raison."[2] In it she develops three theses, which I will present by moving from the general to the particular:

1. The essence of modernity, since Descartes, is nothing but *Ratio,* which can be defined either as the functional rationale of capitalism in its quest for economic profitability (Marxist version), or as the "world of technology" devoting man's energies to the domination

[2] "L'animal, l'homme," *Alliage,* nos. 7/8 (Spring/Summer 1991).

of the earth (Heideggerian version). This vocation, which emerged with Cartesianism, came to fruition with the ideology of the Enlightenment and its belief in progress. Thus there is "an irrevocable ontological complicity between the founding subjectivity and the mechanism," since, once the subject is established as the sole and unique pole of meaning and value, nature can no longer be conceived as anything but a gigantic reservoir of neutral objects, or raw materials destined for human consumption.

2. From this perspective, in which rationality becomes the *absolute* norm of all evaluation, that which is without reason can only be devoid of value. Foucault had already applied a similar grid to his reading of the history of madness, defined as *"déraison,"* unreason. Elisabeth de Fontenay chose another object, the animal: "Because the very essence of this *Aufklärung,* which we translate as *Ratio,* is technology, the Cartesian theory of the animal-machine in its paradoxical self-evidence became the centerpiece of classical science." In the same way one could "deconstruct" the modern representation of the "primitive," the "marginal," or of women, who, according to machinist ideologies, are reputed to be more "intuitive" than "rational." But let's stick to animals . . .

3. From here, technoscience could unabashedly and unreservedly wage a veritable campaign of violence against animals. The proof lies in the close link between experimentation and vivisection, an extreme form of legitimized torture:

> The great affair of our times is thus to force life to give up its secret. "When one speaks of vivisection," writes Georges Canguilhem, "one speaks of the demand that life be maintained as long as possible." Isn't that the very definition of torture? A terrorist theory, a violent vision, an armed eye, an unflinching hand, an implacable collaboration between the scalpel and the microscope: curiosity kills slowly, pleasurably, strong in the knowledge that it is at the service of truth, drunk with consulting so many pulsating breasts. The Enlightenment may have fulfilled rather than abolished the heritage of Christianity by devoting itself to this impassive manipulation of the animal.

Foucault said that "reason is a torture whose subject is the agent." Similarly, Elisabeth de Fontenay speaks of "rational despotism," free of all compassion, of the "work of the *Ratio* under the surveillance of the *Cogito*" for which "nature is nothing but the substratum of domination."

It is unfortunate that the analysis, being too "applied," too unilateral, so obviously underestimates the complexity of the modern universe. As I noted earlier, an entire tradition of "humanitarian" republicanism, from Michelet to Hugo, revolted, *in the name of the Enlightenment,* against the cruelty of men. From a simple empirical point of view, it is most often scientists who, in their professional capacities, are interested in the fate of nature and wish to limit vivisection, or militate among the ranks of the ecologists. To associate contemporary science with Cartesianism results, I'm afraid, from a misconception which the Heideggerian philosophers are the last not to have abandoned. Instead they would have us believe that it is necessary to "deconstruct" humanism at all costs in order to respond to the threat of a metaphysical anthropocentrism. We are given a choice, so to speak, between Descartes on the one side, and Heidegger, Derrida, or Foucault on the other. The strategy is too clunky, the connections too obvious to still be convincing today. We have to get beyond this, we mustn't give in so readily to alternatives that run counter to the diversity of the Enlightenment heritage, which cannot be identified solely with the forces of instrumental or technical reason. Humanism cannot be reduced to the dimensions of a Cartesian "metaphysics of subjectivity" anymore than the rights of man are the "superstructure" of the bourgeoisie.

All the more reason to remain vigilant and avoid the traps humanism sets for itself when it appears to legitimize, in the name of the separation of humankind and the animal kingdom, an imperious domination of the latter. Here is another example worth considering.

## Humanism to the Rescue of Bullfighting?

In an article entitled "L'esprit de la corrida" [The Spirit of Bullfighting], Alain Renaut, who is what I believe one calls an *aficionado,* endeavors to give these "games" an esthetic, if not ethical, justification, in order to facilitate defending their existence against ritual attacks by zoophiles.[3] Right from the start his text has the merit of avoiding the trap of arguing for tradition. To seek to legitimize an institution by the simple fact "that it has existed for a long time" is a senseless enterprise: on these grounds, one could call for the preservation of slavery or *just primae noctis* in countries where these practices have subsisted "uninterrupted" for centuries, to repeat the terms of the law on bullfighting. This seems obvious, yet it musn't be forgotten that it is in the name of this parody of an argument, contrary to all our republican principles, that French legislation continues, to this day, to exempt bullfighting from the laws protecting animals from public cruelty.

Retracing the history of Madrilenian bullfighting, examining and dismissing the sociological and psychoanalytic interpretations of its latent or unconscious meanings one by one, Renaut ultimately concludes that only a philosophical approach can enable one to grasp its true *spirit*. More concerned with *understanding* than with *explaining,* Renaut traces the lines of his own interpretation based on the re-marks of matadors such as Paquirri, the "hero," so it seems of the 1970s, who in 1984 was to fall victim to a little black *toro* named Avispado: "What I really like is educating the beasts. All they know is how to attack. Sometimes slowly, sometimes viciously. They're like children starting school. You have to teach them everything, from b, to a, to ba, to the whole alphabet. You have to discover their potential . . ."

I'm tempted to say he wouldn't be my choice of a babysitter. But let us avoid cheap shots and instead follow our guide in this matter in order to understand this strange confession. Here is what Alain Renaut makes of it:

> Less shocking than naively blundering in its choice of images, Paquirri's admission makes clearly apparent the significance the fight held for a *matador* who conceived of his profession as the will to affirm the superiority of

[3] Alain Renaut, "L'esprit de la corrida," *La Règle du Jeu,* no. 6 (Spring 1992).

reason: because the *toro* represents brute force, because it incarnates all that is not human, bullfighting symbolizes man's combat with nature—a nature that is constantly threatening to engulf him from without and from within, while he attempts to break away from it by countering violence and aggressivity with reason and calculation.

As in Rousseau or Kant, man is defined by this "breaking away" from a natural state to which the animal remains prisoner—and it is the enactment of this difference, the foundation of modern humanism, that we are witnessing at a bullfight. Hence, according to Renaut, its aesthetic quality: it is one of the tangible, visible illustrations of the humanistic idea par excellence, as he indicates in a passage that summarizes the heart of his argument. Here, according to him, is what happens in the arena:

> In barely twenty minutes, a savage force is subjected to the will of man, who manages to slow, channel, and direct its attacks. The submission of brute nature (which is to say violence) to man's free will, the victory of freedom over nature, elicits an aesthetic emotion in the spectator. Not that the goal of tauromachy is to create beauty through elegant and successfully executed maneuvers, but fighting in the bullring means commanding the course of the bull, triggering its direction with one's voice or gesture, altering its trajectory with a simple movement of the wrist, determining as much as possible where the animal will move and where, pass after pass, he will be subjected to the toreador's control: herein resides the true aesthetic dimension of bullfighting, on the basis of which it no longer seems absurd to consider it an art, if it is true that artistic creation is in some way connected to the submission of unreasoning matter to a will that gives it form.

The interpretation is seductive: it has the advantage of furnishing a plausible reading for cultural aspects of a game which does indeed attract more than just sadistic, bloodthirsty crowds. Nonetheless, I cannot share the idea that the love of bullfighting corresponds to a philosophical commitment to humanism. This is not a simple empirical or personal question. It is true that bullfighting interests me very little and that I have some trouble understanding the pleasure others derive from it. But the very fact that I am indifferent means that I am not militantly hostile toward it either. The question lies elsewhere: it has to do with the nature of humanism, an illustration of which Renaut sees or believes he sees in bullfighting. I would like to show in what follows that it still remains Cartesian ("non-Kantian"—just as one says "non-Euclidian") while the object to which it is applied (the living being) is "non-Cartesian."

From a strictly philosophical point of view, the stakes of this debate are far from small: it is a matter of knowing whether a position one claims to be the expression of a "theoretical humanism" implies that nature, in this case the living being, is relegated to the status of an object to be dominated or "civilized." If this were the case, I must admit that the "nonmetaphysical" humanism, whose principles I have attempted to elaborate for the past fifteen years in each of my books, would make little sense to me. I say this because it would mean that there would no longer be a significant difference between the Cartesian goal of making ourselves "masters and possessors of nature" and the Kantian goal of a humanism concerned with respecting *the diversity of the orders of reality.* I say this, too, because the Heideggerian deconstruction of modernity which we have constantly denounced as unilateral would finally be justified: it would rightfully be able to show, in very Cartesian fashion, how humanism *in all its forms* leads to justifying what one would be forced to call a "colonization of nature."

Everyone knows or ultimately finds out that ecology, or at least the ecology movement, possesses some questionable roots, and that the musty smell of the Petainist obsession with one's native soil often lingers in its wake. Must we conclude that the love of man necessarily implies the hatred of nature? Must one give in, because "ecologists can be fascists," to the polemical construction that makes the proof of one's humanity depend on one's degree of disdain toward plants and animals? I don't think so, although I already know, ideological confrontations being what they are, that in the years to come the debate on ecology will increasingly take this caricatural form. Nothing prevents us from anticipating this tendency.

But let's get back to bullfighting. In what way does Renaut's interpretation bring into play a humanism that is still *Cartesian?*

Let us return to the essential element, the element without which the entire argument would make no sense: the fight must take place against a *living* being and not a simple machine. That the bull incarnates brute force, which is "civilized" by the toreador, as Renaut argues, is not in question here (though the idea of "civilization" suggested is somewhat problematic). But clearly it is necessary that this force be neither mechanical, nor entirely natural: no fight is possible against a locomotive (a machine conducted and constructed by men), or against the waves of the ocean (a simple material event). The toreador is not Don Quixote: he must master a living force, which is to say a force that is *mobile, finalized, and, to some extent at least, however small, unpredictable.* Without this condition there could *never* be a victory of the animal over man, and the game would be entirely without interest.

Thus human freedom must prevail, not over brute force in general but over a *living being.* I maintain that, given these conditions, the esthetic satisfaction derived from the spectacle of the bullfight cannot be based on the fact that the latter illustrates or "enacts" a nonmetaphysical idea of humanism. It is clear, in effect, that this type of domination of the living being, which is perfectly satisfying from a Cartesian perspective, is unacceptable within the broader Kantian perspective which Renaut in fact claims as his own. Not at all out of "sentimentality," as fools might say, but because, for anti-Cartesian philosophical reasons which must be examined, *the reduction of an animal to the state of a thing* (its killing) cannot be the object of a *game.* It may, in some instances, be a necessity, but never entertainment for those who are attentive to the *diversity of the orders of reality.* And no one can pretend that the "killing" (or even the animal's suffering) is not essential to bullfighting, for as we know it was on this precise point that the legislation, which since the mid-nineteenth century had forbidden fighting with or among animals, had to be changed, following long and heated debates.

We know Kant's own position: animals have no rights (as zoophiles would have it), but on the other hand we do have certain, indirect duties toward them, or at least "on their behalf" ("*in Ansehung von*," says Kant). The justification for this "behalf" may be deemed insufficient. Why would there be duties "on behalf" of animals if there were not *within them* some *intrinsic* particularity worthy of respect? Kant suggests a path for reflection when he writes the following: "Because animals are an *analogue* of humanity, we observe duties toward humanity when we regard them as analogous to humanity, and thus we satisfy our obligations toward it." Why? Simply because, as opposed to what Descartes and his automata makers thought, the living being is not a thing, the animal is not a stone, nor even a plant. So what, one may ask? So life, defined as "the faculty to act according to the representation of a goal," is an *analogue of freedom. As such* (that is to say in its highest forms) and because it maintains a relationship of analogy with that which makes us human, it is (or should be) the object of a *certain* respect, a respect which, by way of animals, we *also* pay ourselves.

It is not by chance if Kant links up here with one of Judaism's most profound intuitions: man is an antinatural being, a being who lives by law (this, in fact, is what prohibits both the Kantian tradition and Judaism from identifying with radical ecology). He can thus dispose of plants and animals *to a certain degree*—but not at will (*nach Belieben*), not by *playing* at killing them, even if within the rules of art and as testimony to his humanity. According to the Pentateuch, slaughtering is to be practiced not only without cruelty but in moderation. There is great wisdom and depth here, *for this position is not accompanied by the "naturalist" and vitalist principles that ordinarily justify zoophile arguments.* There is no possible confusion here between man and animal in the great cosmic order. No reduction, either, of the dignity of one or the other to the simple calculating logic of pleasure and pain. Instead, attention is paid to the enigmatic uniqueness of animals which the Cartesian machinery, entirely devoted to the domination of the earth, casts unreservedly on the side of things.

This friendly disagreement brings three further reflections to mind.

The first is that the *philosophical* status of the living being apparently has yet to be considered, stymied as we still are by the powerful antinomy between the (Cartesian) mechanism and (romantic, then

Nietzschien) vitalism. Rousseau, Kant, and Fichte have pointed toward paths for reflection. As we have seen, they never reduced animals to simple mechanisms. When Rousseau described animals as being governed by the code of instinct, he granted that they enjoyed the privilege of affectivity and even of thought in the same breath. In a course during the winter semester of 1929–1930, the "first" Heidegger, who was closer to Kant than is ordinarily believed, had also attempted to elucidate the meaning of animal ambiguity through his analysis of the proposition: "The stone is without world, the animal is poor in world, man is creator of world (*Weltbildend*)." "Poor in world" means it is "less" than man, no doubt, who can "exist," *transcend* natural cycles to rise to the consideration of questions that no longer bear only on intraworldly facts but on the world's existence itself. It is nonetheless "more" than the stone, which attains no representation and is endowed with no finalized movement. However, Heidegger emphasizes that the terms "more" and "less" are misleading here: they suggest a continuity between orders of reality, whereas these orders are qualitatively different. Which doesn't mean that no *analogies* can be made between them, in particular in the case of the latter two: let's not forget that many things have been done against/to animals, *exactly as many things have been done against/to humans*. Here the expression "it's not by chance . . ." is appropriate.

Faced with the reproach to the humanist tradition, that it is nothing but anthropocentrism, a "metaphysics of subjectivity," the verification of which can be seen in the devastation of the earth and the torture of animals at the heart of a "technological world," it is time to show that a non-Cartesian humanism evades the absurd alternative to which radical ecology condemns us: the "step backward" or barbarity. And since the philosopher is occasionally asked to elaborate on the concrete consequences of the distinctions he establishes, here is one that clearly indicates the dividing line with Cartesianism: I am speaking of what takes place in Canada, where "animal ethics committees," working in university hospitals, are responsible for monitoring, and if need be modifying, the protocol of experiments in which animal suffering is involved. Why, after all, should we passively accept the right of irresponsible scientists to indulge in the most improbable experiments in the privacy of their laboratories? I am not claiming that this is the general rule—though every year countless experiments are performed in which great suffering is inflicted on thousands, even millions of animals, without the slightest benefit. And yet rules of deontology are no doubt already in effect among most scientists. I see nothing wrong, however, with the idea of ensuring that these rules are observed in practice, and that one not be allowed to do anything at all, not only "on behalf" of animals but to them.

I had an opportunity to closely examine the report of the meetings of the University of Quebec's animal ethics committee: I must admit that in several hundred pages, I saw only intelligent remarks and useful advice. All things considered, once I got over my initial reactions I became convinced that there was nothing ridiculous about prescribing a certain type of analgesic or insisting on a less painful manner of killing an animal to be dissected—rather than considering that there is no matter for reflection here and that anything and everything should be allowed.

≈

One may object that respect for animals is a "projection" in the psychoanalytic sense of the word—which is why it is characteristic of children. In a sense, this is true. Still it should be noted that if man cannot help but see himself, to some extent, in the enigmatic nature of *animals,* it is not only due to a psychological projection but also to a *philosophical* one, because the *analogue* of freedom can never leave entirely cold one whom the eighteenth century would have called "a man of taste." Freud said that the latter must renounce, even if regretfully, the pleasure of puns. To which I will add the enjoyment of bullfights and other games of that order. What we need is a synthesis between the Heideggerian *Gelassenheit,* or "letting be," and the imperious "civilizing" activity of the Cartesians. For perhaps the circumscribed respect we owe animals, far from being inscribed in nature or a burden placed upon us by civilization, is in this sense a matter of *politeness* and *civility.*

*part two*

## THE SHADOWS
## OF THE EARTH

## "Think Like a Mountain":
## The Master Plan of "Deep Ecology"

"Think like a mountain": the task promises to be a bit tricky for some. In any case, it is in these terms that Aldo Leopold, whom many consider to be the father of "deep ecology," invites us to overturn the paradigms that dominate Western societies. Quoted a thousand times in American literature, the preface of his essay on *A Land Ethic* develops the major theme of this strange revolution:

> When God-like Odysseus returned from the wars in Troy, he hanged all on one rope a dozen slave-girls of his household whom he suspected of misbehavior during his absence. This hanging involved no question of propriety. The girls were property. The disposal of property was then, as now, a matter of expediency, not of right and wrong. Concepts of right and wrong were not lacking from Odysseus' Greece . . . There is as yet no ethic dealing with man's relation to land and to the animals and plants which grow upon it. Land, like Odysseus' slave-girls, is still property. The land-relation is still strictly economic, entailing privileges but not obligations.[1]

The conclusion is clear: after having succeeded in rejecting the institution of slavery, we need to go one step further, to finally take nature seriously and consider it as endowed with *intrinsic* value

[1] Aldo Leopold, who died in 1948, remains one of the most prominent figures of American ecology. His major book, *A Sand County Almanac* (a collection of essays including *A Land Ethic,* his most famous, published in 1949), had an inestimable influence on the deep ecology movement.

worthy of respect. This conversion—the religious metaphor is not unfounded here—presupposes a deconstruction of "human chauvinism," the root of anthropocentric prejudice par excellence, a prejudice that leads us to consider the universe as the stage for our actions, the mere periphery of a center which we have instituted as the sole subject of value and rights.

Hence the debate that divides American ecology and which is now tending, via Germany in particular, to make headway in Europe: should we merely be safeguarding the sites where *we* live because their deterioration might affect *us,* or, on the contrary, should we be protecting nature in and of itself, because we are discovering that it is not simply a collection of raw materials, endlessly pliable and exploitable, but a harmonious and fragile system, in itself more important and wondrous than the ultimately tiny segment that constitutes human life? Clearly, these two positions may sometimes meet in practice, in combating a particular industrial threat, for instance. But fundamentally, when it comes to the philosophical and political principles they bring into play, the two are diametrically opposed: the first preserves the heritage of modern humanism intact (it is in man's interest to respect the earth), while the second implies the most radical questioning of it. Humanism is not the answer to the crisis of the modern industrial world but rather an original sin, the primary cause of evil. Hence the diehard nature of a conflict which Bill Devall, one of the main theoreticians of this new fundamentalism, presents in these terms:

> There are two great streams of environmentalism in the latter half of the twentieth century. One stream is reformist, attempting to control some of the worst of the air and water pollution and inefficient land use practices in industrialized nations and to save a few of the remaining pieces of wildlands as "designed wilderness areas." The other stream supports many of the reformist goals but is revolutionary, seeking a new metaphysics, epistemology, cosmology, and environmental ethics of the person/planet.[2]

[2] Bill Devall, "The Deep Ecology Movement," *Natural Resources Journal* 20, no. 2 (April 1980).

It is this as yet unprecedented vision of the world that Bill Devall, following the lead of the Norwegian philosopher Arne Naess, who was the first to present its "ideal-type,"[3] proposes to name *deep ecology*. It would be wrong to imagine that we are dealing with a simple curiosity, an exotic symptom of the folly that seems to overtake American university professors on occasion, as when they succumb to the fashion of "deconstructionism" or to the imperative of "political correctness." Deep ecology finds support outside the academic community as well as abroad: it has inspired the ideology of movements such as *Greenpeace* or *Earth First,* for instance, of powerful associations like the Sierra Club, but also of a significant portion of Green parties, and to a large extent the work of popular philosophers such as Hans Jonas or Michel Serres.

It is, therefore, necessary to take stock of such an enterprise. Though it may seem strange at first glance (and even, perhaps, at

[3] See Arne Naess, "The Shallow and the Deep, Long-Range Ecology Movement: A Summary," *Inquiry* 16 (1973). See also, by the same author, "The Deep Ecological Movement: Some Philosophical Aspects," *Philosophical Inquiry* 8 (1986). The literature devoted to this movement is voluminous, but also, it must be said, very repetitive. For an intitial overview, I suggest to the reader interested in learning more about this topic the following articles, which seem to me among the most significant: George Sessions, "The Deep Ecology Movement: A Review," *Environmental Review*, no. 9 (1987)–(as its title indicates, Sessions reviews the principal ideas and major books in the history of deep ecology); Richard and Val Routley (who is none other than Val Plumwood, the theoretician of ecofeminism), "Against the Inevitability of Human Chauvinism," in K. E. Goodpaster and K. Sayre, *Ethics and the Problems of the 21st Century* (Notre Dame, Ind.: University of Notre Dame Press, 1979); J. Baird Callicott, "Non-Anthropocentric Value Theory and Environmental Ethics," *American Philosophical Quarterly* 21 (October 1984); Michael E. Zimmerman, "Toward a Heideggerian Ethos for Radical Environmentalism," *Environmental Ethics* 5 (Summer 1983); Paul W. Taylor, "The Ethics of Respect for Nature," *Environmental Ethics* 3 (1981); Roderick Nash, *The Rights of Nature: A History of Environmental Ethics,* (Madison: *University of Wisconsin Press,* 1989). For a critical reflection on the movement, see also Luc Béjin, "La nature comme sujet de droit? Réflexions sur deux approches du problème," *Dialogue* 30 (1991). I should add that in many respects the concerns of deep ecology are also those of what Illitch called "radical" ecology.

second), it nonetheless possesses a systematic coherence sufficiently impressive to seduce many of those who have been left stranded by the political void and the end of utopian visions. I have already mentioned how Christopher Stone, in the trial brought by the Sierra Club against the Walt Disney company, proposed pleading in favor of a legal status for trees and valleys. Before addressing the most general principles it will no doubt be useful, given the unusual nature of the project, to look at two concrete examples of the manner in which deep ecologists mean to renew our ethical-legal approach to nature, in the wake of "the death of man" and the deconstruction of anthropocentrism.

## Duties Concerning Islands

In an essay bearing this title, Mary Midgley, a British philosopher and author of numerous works on ecology and animal rights, proposes a new version of *Robinson Crusoe,* whose journal, once revised and edited, would reveal the following events:

> 19 Sept. 1685. This day I set aside to devastate my island. My pinnance being now ready on the shore, and all things prepared for my departure, Friday's people also expecting me, and the wind blowing fresh away from my little harbour, I had a mind to see how all would burn. So then, setting sparks and powder craftily among certain dry spinneys which I had chosen, I soon had it ablaze, nor was there left, by the next dawn, any green stick among the ruins. . . .[4]

Here is Midgley's commentary:

> Now, work on the style how you will, you cannot make that into a convincing paragraph. Crusoe was not the most scrupulous of men, but he would have felt an invincible objection to this senseless destruction. So would the rest of us. Yet the language of our moral tradition has tended strongly,

[4] Mary Midgley, "Duties Concerning Islands," in *Environmental Policy: A Collection of Readings* (University Park and London: Pennsylvania State University Press, 1983.

ever since the Enlightenment, to make that objection unstateable.[5]

The argument's meaning is twofold: on the one hand, it is a matter of *showing* that there exist incontestable duties toward non-human beings. I emphasize the word "showing" because Midgley is not really reasoning here but rather invoking something which, according to her, every human being in good faith should encounter in his conscience upon hearing of Robinson's delirium. And this sentiment of horror—here is the second stage—can no longer find the words to express itself today, the moral tradition of humanism, since the Renaissance, having conceived of our ethical duties on the *contractualist* model of commitment toward other individuals, considered as equals—"The theoretical model which has spread blight in this area is, of course, that of the social contract, and, to suit it, that whole cluster of essential moral terms—right, duty, justice and the rest—has been progressively narrowed"[6]—*the field of validity of these concepts having been limited solely to human beings.* Robinson's anecdote makes manifest, at least according to Midgley, the existence of duties toward entities other than men. And not just toward islands. There exist moral obligations toward "the dead, posterity, children, the insane, embryos, human and otherwise, sentient and nonsentient animals, plants of all kinds, artefacts, inanimate but structured, ecosystems, families, species, countries, the biosphere, oneself, and God!"

This is one way of saying that humanist morality has drastically reduced the notion of duty! But it also suggests the extent to which deep ecology, with such a broad concept of obligation, could become a moralizing power of the first order were even a relatively weak dose of social control to give it real power over individuals. This is the case today in the United States and in Germany. No one can say what lies in store for the rest of Europe.

From a philosophical point of view, the entire tradition of humanism inherited from the doctrines of the social contract and the rights of man must be deconstructed in order to create a legal status

[5] Ibid.
[6] Ibid.

for islands and forests. As Midgley insists, "We have quite simply got many kinds of duties to animals, to plants, and to the biosphere. But to speak in this way we must free the term once and for all from its restrictive contractual use, or irrelevant doubts will still haunt us." Thus, the question of the ethical recognition of inanimate beings tends to become the criteria for the success or failure of such a deconstruction of modernity, as Roderick Nash, another theoretician of the movement, suggests:

> Do rocks have rights? If the time comes when to
> any considerable group of us such a question is no longer
> ridiculous, we may be on the verge of a change in value
> structures that will make possible measures to cope with
> the growing ecological crisis. One hopes there is enough
> time left.[7]

Nash's pessimism is unjustified: the question of the rights of things is far from a dead letter or a simple philosophical fiction, as evidenced by the vigor of the legal debates dealing with the question of "crimes against the ecosphere."

## Crimes against the Ecosphere

In 1985, the very official Law Reform Commission of Canada, founded in 1971 with an eye toward modernizing federal legislation, published a working paper entitled "Crimes against the Environment."[8] Among the numerous propositions destined to be discussed in Parliament, one recommends adding to the criminal code a new and specific crime concerning acts that "seriously compromise a fundamental societal value and right, that of a safe environment or the right to a reasonable level of environmental quality."

Yet the members of the Commission, faithful to the essence of Anglo-Saxon law, which aims to protect identifiable interests, admitted their inability to consider nature itself a legal subject: "The scope of a Criminal Code offence against the environment should not

[7] See Roderick Nash, "Do Rocks Have Rights?" 10 *Center Magazine* 2 (1977).

[8] "Crimes against the Environment," Working paper no. 44, 1985.

extend to protecting the environment for its own sake apart from human values, rights and interests." Although determined to advance the ecologist theses, the commission chose the environmentalist camp over that of deep ecology. It stayed within the framework of classical humanism, thus of the anthropocentrism so decried by the radicals: "The present Criminal Code in effect prohibits offences against persons and property. It does not, in any explicit manner, prohibit offences against the natural environment itself."

Despite the distance maintained from fundamentalist themes, two elements nonetheless belie the latter's progress in "enlightened" opinion: first, the title of the report itself, which is highly ambiguous, since it seems to concede the idea that there can be such a thing as a crime against nature—and not only against men. Second, it is clear that the idea of a right of objects is sufficiently in the air for jurists called upon to make propositions likely to come before Parliament to feel the need to discuss it explicitly. Despite their reservations, after extensive arguments the commission admitted that some serious pollution can be considered an authentic crime, in the legal sense of the word.

The reaction of deep ecology circles was nonetheless quite animated, as evidenced by the debates that followed, and was reflected, among other places, in an article by Stan Rowe entitled "Crimes against the Ecosphere."[9] His conclusions are worth reporting here insofar as they are exemplary of antireformist positions.

> [The Report entitled] *Crimes against the Environment* accepts the traditional anthropocentric (homocentric) position that takes *environment* to be exactly what its etymology suggests: the context and surroundings of things of greater importance—namely, people. In this popular sense *environment* is peripheral; the concept is its own pejorative. Logically then, environment's defence must be restricted to terms of human utility; 'it is a societal valve and right,' not a thing of inherent worth. I argue that the alternative—recognition of

[9]Stan Rowe, "Crimes against the Ecosphere," in R. Bradley and S. Duguid, *Environmental Ethics,* vol. 2 (Burnaby, B. C., Canada: Simon Fraser University, 1989).

environment's intrinsic values and thereby its inherent rights—provides the only incontrovertible basis for protecting it against crimes of despoliation and degradation.

The rest of the text develops two ideas, both also perfectly representative of radical principles: that of the sacred nature of universal life, of the "biosphere," and that of the disastrous consequences of the *Declaration of the Rights of Man and of the Citizen* and of the associated humanism. As for the first point, Rowe takes great pains to stipulate that it is not primarily a question of human life but of the ecosphere as a whole. A strange hierarchy but one that results from the so-called biospheric egalitarianism principle, according to which it is fitting to protect the whole before its parts. *Holism,* that is, the philosophical thesis according to which the totality is morally superior to individuals, is thus explicitly embraced as a positive theme of deep ecology. Countering the individualism characteristic of Western modernity, the word itself must be reinstated as a positive term since "the ecological system, the ecosphere, is the reality of which men are only one part. They are embedded in it and totally dependent on it. Here is the scientific source of the environment's *intrinsic* value." It is not surprising, then, that the critique should extend into a spirited denunciation of the ideals of the French revolution:

> The French *Déclaration des Droits de l'Homme et du Citoyen* defined liberty as "being unrestrained in doing anything [presumably to the natural world] that does not interfere with another's rights." In line with this popular sentiment . . . George Grant defined liberalism as the set of beliefs proceeding from the central assumption that man's essence is his freedom, and therefore what chiefly concerns man in this life is *to shape the world as he wants it.* Here is the prescription for the massive environmental destruction that is evident wherever western culture's influence is felt; destruction whose motivation only the recognition of nature's intrinsic values and rights can overcome.

Speaking against the federal commission, Rowe, therefore, proposes that "the ecosphere ought to be valued above people" and that by analogy with the concept of "crimes against humanity," the notion of "crimes against the ecosphere" should be elaborated, among which would prominently figure "fecundity and exploitive economic growth, both encouraged by the homocentric philosophy." I will leave out the contents of the thesis here (but one day radical ecologists will have to recognize that the birth rate is lower in Europe and the United States than in the Third World, while environmental concerns are infinitely more developed: the modern world isn't all negative!). What must be retained at the very least is that deep ecology has made *holism* and *antihumanism* overt slogans in its fight against modernity (the terms themselves, considered positive, are, I repeat, omnipresent in its literature). This point should be emphasized. It enables us to understand why this form of fundamentalism has come to be politically "unclassifiable," with its combination of themes that traditionally belong both to the extreme Right and to the extreme Left. The reason for this is fairly clear: as I suggested earlier, the external critique of modernity can only occur in the name of a radically different other world, situated somewhere either prior to or after the vilified present (depending on whether the perspective adopted is one of neoconservative romantic nostalgia or a belief in the radiant future of progressive utopias).

Hence the ambiguity of an ideology that constantly lends itself to a double reading. Arne Naess and George Sessions have attempted to gather the principle themes in a text which we will quote here in its entirety, since it is one of the most reliable manifestos of the movement, explaining, as it says, "the key terms and phrases . . . tentatively proposed as basic to deep ecology":

1) The well-being and flourishing of human and non-human Life on earth have value in themselves (synonyms: intrinsic value, inherent value). These values are independent of the usefulness of the non-human world for human purposes.

2) Richness and diversity of life forms contribute to the realization of these values and are also values in themselves.

3) Humans have no right to reduce this richness and diversity except to satisfy vital needs.

4) The flourishing of human life and cultures is compatible with a substantial decrease of the human population. The flourishing of non-human life requires such a decrease.

5) Present human interference with the non-human world is excessive, and the situation is rapidly worsening.

6) Policies must therefore be changed. These policies affect basic economic, technological, and ideological structures. The resulting state of affairs will be deeply different from the present.

7) The ideological change is mainly that of appreciating life quality (dwelling in situations of inherent value) rather than adhering to an increasingly higher standard of living. There will be a profound awareness of the difference between big and great.

8) Those who subscribe to the foregoing points have an obligation directly or indirectly to try to implement the necessary changes.[10]

The last remark ties in with the theoretical aspects of the program and its practical desire to found a new militant movement. Without questioning the self-definition of deep ecology by two of its representatives, it seems to me possible, by taking some distance, to in turn propose my own model for it that brings other information into the picture and attempts to situate it within the logic of a deconstruction of modernity.

*The Critique of "Western Civilization": Revolution versus Reformism*
Let's clear up one misunderstood point: certain deep ecologists, including Roderick Nash and, to a certain extent, Stone himself, wanted to inscribe the recognition of the rights of nature within the logic of democratic societies. As in the utilitarian tradition that we have seen in operation in the call for animal rights, it would be a matter of showing that after the emancipation of blacks, women,

[10] Naesse, "The Deep Ecological Movement: Some Philosophical Aspects," p. 14.

children, and animals, the time has come for trees and rocks. According to this scenario, a nonanthropocentric relationship to nature fits in with the general and continuous movement toward liberation that characterizes the United States. This presentation of things is fallacious (an attempt to confer respectability on deep ecology by introducing it into the dynamic of American society). Indeed, it is clear that the idea of an intrinsic right of beings in nature is in radical opposition to the legal humanism that dominates the modern liberal universe. Most deep ecologists were correct, in fact, in considering their own project as belonging within the orbit of what the 1970s termed a "counterculture" with respect to the dominant Western model. Bill Devall clearly emphasizes this:

> Deep ecology, unlike reform environmentalism, is not just a pragmatic, short-term social movement with a goal like stopping nuclear power or cleaning up the waterways. Deep ecology first attempts to question and present alternatives to conventional ways of thinking in the modern West.[11]

It is a matter, of course, of specifying exactly what is being targeted in this "modern West" and in the name of which principles the critique is being launched. Though given impressionistically, with broad strokes, the response nonetheless forms a figurative image. The following are denounced, in order of their appearance in history: the "Judeo-Christian tradition," because it places the spirit and its law above nature, and Platonic dualism, for the same reason; the technical concept of science that triumphed in Europe beginning in the seventeenth century with Bacon and Descartes, for it reduces the universe to a warehouse of objects to serve man; and the entire modern industrial world, which gives priority to the economy over all other considerations. One cannot change the system by merely refitting it, as reformists naively believe: What is needed is a true revolution, including on an economic level, which implies that the critique of the modern world draws sustenance from radical principles.

[11] Bill Devall, "The Deep Ecology Movement," *Natural Resources Journal* 20 (April 1980).

Thus the sources of deep ecology are also situated radically outside of Western civilization. We find jumbled together a strong reference to the misunderstood values of the East, which young Americans discovered in the 1950s and 1960s through "marginal" books like those on Zen Buddhism by Alan Watts or Daisetz Suzuki.[12] Similarly, the guilt-laden reevalutaion of the American Indians' traditional ways of life has also furnished "alternative" models, their religious traditions and customs providing an example of a life in harmony with nature in its original form. Here too, the 1960s offer a pleiad of heralds, beginning with Carlos Castaneda, whose entire work is reputed to show the superiority of "ancient wisdom" over the folly of contemporary technology. But thinkers like Marcuse, Ellul, and especially Heidegger are also brought to the stand as witnesses against the West, while among classical philosophers, Spinoza has been posthumously reinstated against the ignoble Descartes, founding father of modern anthropocentrism. For has the author of *Ethics* not shown, by virtue of his pantheism, that nature is divine, and as such endowed with intrinsic value, and that man, far from being its master and possessor, constitutes but an infinitesimal part of it? It is from this perspective that Robinson Jeffers, a California philosopher and radical Spinozaist who inspired the work of deep ecologists such as George Sessions,[13] explicitly calls for the creation of an "inhumanist" philosophy, which alone would be able, in his view, to reverse the dominant paradigm of anthropocentrism and finally grant nature the rights it deserves.

*Antihumanism or the "Preference for Nature"*

It is "humanist hostility" toward "things nonhuman" that explains the fact that "Western culture differs from most others in the breadth of destructive licence which it allows itself, and, since the seventeenth century, that licence has been greatly extended"[14] The theme

[12] A Watts, *The Spirit of Zen: A Way of Life, Work and Art in the Far East* (1955); D. Suzuki, *Essays in Zen Buddhism* (1961).

[13] See George Sessions, "Spinoza and Jeffers on Man in Nature," *Inquiry* 20 (1977).

[14] Midgley, "Duties Concerning Islands."

has been inexhaustibly repeated, even in best-selling books such as those by David Ehrenfeld, *The Arrogance of Humanism* (1979), or John Lovelock, *Gaia* (1979). Today it even has its French translation in *Le contrat naturel* by Michel Serres. Since Descartes and his formidable plans for control, we have continually and unrestrainedly dominated the world. First, we deprived it of its mystery by declaring it manipulable and calculable at will. Gone are the days of animism and "occult qualities," those mysterious forces with which the nature studied by medieval alchemists was still imbued. But there is more: not content to merely rob the universe of its enchantment, with the birth of modern industry we have established the means to consume it to the point of depletion. This, according to Serres, is what is new to the affair: for the first time, no doubt, in the history of humanity, the problems posed by the devastation of the earth have become global. As on a ship lost in a storm, there is no escape possible, nowhere to seek salvation and protection. The world we have treated as an *object* has become a *subject* again, capable of revenge: worn out, polluted, mistreated, it now threatens to dominate us in turn. Hence the idea of a "natural contract," analogous to the eighteenth-century philosophers' social contract: just as the latter proposed to govern relations among men through law, now we must envisage our relationship to nature under the same auspices. More concretely, perhaps, entering into a contract with nature would mean reestablishing a certain justice. Man's relationship with nature, now one-directional and inegalitarian, must go from "parasitic" to "symbiotic," accepting the notion of giving back what one borrows:

> Let's get back to nature! This means, in addition to the exclusively social contract, signing a natural contract based on symbiosis and reciprocity in which our relationship to things would leave behind mastery and possession in favor of appreciative listening. The law of mastery and property boils down to parasitism. On the contrary, the law of symbiosis is defined by reciprocity: as much as nature gives to man, man must return to nature, *a new legal subject.*[15]

[15] Michael Serres, *Le contrat naturel* (Paris: Flammarion, 1990), p. 67.

That such a program implies a radical questioning of the humanist tradition, even a certain return to the old conceptions of law, and not a simple addition to the social contract, Michel Serres himself cannot fail to emphasize. Like the American deep ecologists, he is obliged to put the French Declaration of 1789 on trail, a declaration that "ignores and glosses over the world" to the point of making it its victim. According to the definitions of law it carries, man alone, "the subject of knowledge and action, enjoys all rights while its objects enjoy none . . . This is why we are necessarily fating the things of this world to destruction." Therefore, we must reverse the Declaration's humanist perspective. Instead Serres would have us adopt the point of view of the objects: "Once again, we must give a ruling on the vanquished, by writing the law of beings who have none."

One may object, and not without reason, that this is a metaphorical fable more than a rigorous argument. It seems rather difficult, indeed, to confer literal meaning on the contract proposed by Serres ("Hello Mother Nature, let's be friends"). In his *The Imperative of Responsibility,* Hans Jonas also establishes a strict philosophical link between the need for a critique of humanism and that of a recognition of the rights of nature. The chapter entitled, significantly, "Has Nature 'Rights' also?" responds without hesitation in the affirmative. Referring to the opinion according to which "our duties extend farther" than to "the interest of man alone," he decrees that "the anthropocentric confinement of former ethics no longer hold." Yet Jonas does not push the analogy between humankind and nature to the point of considering nature a "person" in the classical sense of the word: indeed, it cannot *contract* commitments with us—which is why Jonas considers somewhat incoherent and forced the idea of a "natural contract."[16] Nonetheless, according to him, "it is at least not senseless anymore to ask whether the condition of extrahuman nature, the biosphere as a whole and in its parts, now subject to our power, has become a human trust and has something of a moral claim on us not only for our ulterior sake but for its own and in its own right. If this were the case it would require quite some rethinking in basic prin-

[16] See "De la gnose au 'principe responsabilité': un entretien avec Hans Jonas," *Esprit* (May 1991), p. 15.

ciples of ethics. It would mean to seek not only the human good but also the good of things extrahuman, that is, to extend the recognition of 'ends in themselves' beyond the sphere of man and make the human good include the care for them"—which, according to *The Imperative of Responsibility,* no moral humanist has yet been able to do, and for very good reason.

If we leave aside the contractualist metaphor, which in effect is not very rigorous in the context of reinstating the rights of nature and thus opposing the anthropocentrist logic inherent to the notion of the contract (this was, the reader may recall, Mary Midgley's argument), we notice how similar the perspectives of the American, German, and French versions of deep ecology are: in all cases, it is a matter of questioning the modern tradition of legal humanism to arrive at the idea that nature possesses an *intrinsic value* and that it is, as such, worthy of respect. It is with this in mind that Jonas even proposes to extend the concept of "end in itself," which, as we know, was a concept that Enlightenment thought was intent on reserving exclusively for humans, natural objects possessing only the status of means. In the process of this reevaluation, the entire *Cosmos* may well be assigned a positive coefficient higher than that of humankind itself, since in the hierarchy of beings it constitutes the primary condition: nature can do without men, but not vice versa, which is why the idea of a "preference for nature" finds itself gradually legitimized as all in all the most logical metaphysical horizon of deep ecology.

This too is the ultimate meaning of the reference to Spinoza. As German ecologist Claus Meyer-Abisch writes, drawing inspiration from this pantheistic metaphysics: "Nature continues within us insofar as it becomes language and art, and in other living beings insofar as the latter live, so to speak, their lives (i.e., the life of nature). Our life and that of the surrounding world is *its* life." The *"Natura naturans,* the creative force, is itself everything everywhere. It is thus the true center of the world." The conclusion, which seeks to reinstate holism, logically follows: "Whether a tree dies or whether a man dies, in either case a living being dies and returns to the earth."[17]

[17] K. M. Meyer-Abisch, *Wege zum Frieden mit der Natur* (Munich, 1984), pp. 90, 100, 187.

Indeed, but can we be sure that the two losses have the same meaning, the same value, that the destruction of entire populations is really comparable to that of trees in our forests?

Not that the "inhumanist" theses of fundamental ecology appear only in the upper reaches of professional philosophy. We find them more or less clearly formulated in all the Green movements in Europe, as evidenced, for example, by this passage from the most recent book by Antoine Waechter:

> The word "nature" is expurgated from all discourse as if it were indecent, or at the very least childish, to evoke what it designates. The word environment, apparently more credible, has won out . . . The choice is significant, Etymologically, the word "environment" designates that which surrounds, and in this context, more precisely, what surrounds the human species. This anthropocentrist vision conforms to the spirit of our civilization of conquest, whose only reference point is man and whose every action tends toward total mastery of the earth . . . This notion is one of the fundamental points of difference with ecologist philosophy, which perceives the human being as one organism among millions of others and considers that all forms of life have a right to an autonomous existence.[18]

This theme also constitutes the ideological foundation of an organization such as *Greenpeace,* which states its position clearly in an April 1979 editorial published in its *Chronicles:*

> Humanist value systems must be replaced by suprahumanistic values that bring all plant and animal life into the sphere of legal and ethical consideration. And in the long run, whether anyone likes it or not, force will eventually have to be brought to bear against those who would continue to desecrate the environment.

The warning is clear: the bypassing of humanism in favor of the vegetable and animal kingdom in matters of ethics and law will not

[18] Antoine Waechter, *Dessine-moi une planète* (Paris: Editions Albin Michel, 1990), p. 151.

occur without coercion—an argument that is, in fact, coherent from the point of view that says it is time to put an end to the logic of the infamous "rights of man," which have served for little other than the legitimization of forgetfulness at best, the destruction of the world through the unleashing of technology at worst.

Similarly, someone like Jean Brière, a former member of the Greens and a close friend of Waechter's, suggests "stemming at the source the overproduction of children in the third world," while Jean Fréchaut, also a member of the Greens at one point, dreams of a "global government that can subjugate populations in order to reduce pollution and alter desires and behaviors through psychological manipulation"![19] Are these thoughts representative only of fringe elements, who are not expressing the general, pacifist, sentiments of radical ecologists? Perhaps, probably even, if we only look at what they say, which is scandalous indeed. But deep down? Are they not saying out loud what others think to themselves? What exactly does one mean when one speaks of the "population bomb"? No one can argue that there is not a real problem and that the birth rate in certain countries may worry those responsible, in one capacity or another, for the fate of their populations. But as always there is more than one way of approaching a problem: when we get to the point of arguing that the ideal number of humans, *from the point of view of nonhumans,* would be 500 million (James Lovelock), or 100 million (Arne Naess), I would like to know how one plans to realize this highly philanthropic objective. For, here too, the dreams of radical ecologists often turn to nightmares, as in the case of the death program evoked by William Aiken and published in an anthology that enjoys an excellent reputation: "In fact, massive human die backs would be good. Is it our duty to cause them? Is it our species' duty, relative to the whole, to eliminate 90 percent of our numbers?"[20]

Let us not forget that this type of thinking led many German ecologists to prefer the Soviet system to liberal democrat regimes.

[19] These two passages are quoted and commented upon by Pronier and Le Seigneur, *Génération verte* (Paris: Presses de la renaissance, 1992), pp. 191, 208.

[20] William Aiken, in *Earthbound: New Introductory Essays in Environmental Ethics,* edited by Tom Regan (New York: Random House, 1984), p. 269.

Jonas himself, whom some consider to be an authentic philosopher, provided an example when, in the late 1970s, he still considered that totalitarianism had the "merit" of rigorously planning consumption and of thereby constraining its fortunate subjects to a "healthy frugality." Let me refrain from irony, which comes easily in this context. Let's just say that the pages Jonas devotes to this subject in his *Imperative of Responsibility* add nothing to the glory of this century's intellectuals and in large part invalidate the very title of his book.[21]

*Frankenstein and The Sorcerer's Apprentice: Technology in Question*
Is deep ecology hostile to "science"? Formulated in this manner, the question makes little sense. The terms must be clarified. If by science we mean global wisdom, a new cosmology that takes as its sources the Ancients' (preferably Greek, Chinese or Indian) traditional and religious visions of the world, the answer is no. If instead we are talking about modern technology, closely tied in with the advent of anthropocentric Western civilization, entirely oriented toward production and consumption, it is clear that, from the fundamentalist point of view, the answer can only be yes. An emphatic yes, even, to the point that those philosophers who made the critique of technology central to their works have become required reading—beginning, of course, with Heidegger, who is cited by Bill Devall as being, along with Whitehead, the "most influential" European thinker on the movement.[22]

It is in large part on this same theme that *The Imperative of Responsibility* built its formidable success. From the first page of the preface to the French edition, the cards are on the table:

The preliminary thesis of this book is that the promise of modern technology has reversed itself into a threat . . . The

[21] Dominique Bourg is one of the few to have correctly perceived this aspect of Jonas's book. See her excellent articles on this subject published in the journal *Esprit*.

[22] Devall: "In particular, more American philosophers, both those with an interest in ecological consciousness and those interested in contemporary philosophers, are discussing Heidegger's critique of Western philosophy and contemporary Western societies."

subjugation of nature with a view toward man's happiness
has brought about, by the disproportion of its success,
which now extends to the nature of man himself [Jonas is
thinking here of the life sciences, and of the possibility of
performing genetic manipulations on the human species],
the greatest challenge for the human that his own deeds
have ever entailed.

Fear of technology brings a return of the old science fiction
myths: in the story of Frankenstein as in that of the Sorcerer's Apprentice, we have a reversal by which the creature becomes its master's
master. Deep ecology likes to apply the metaphor to technology.
According to Jonas, the history of our relationship to the world has
occurred in three stages, each characterized by a certain type of
power. First the power man progressively conquered from nature.
This corresponds to the emergence of technology as part of a plan to
dominate the earth. But today this first type of power is tending to
reverse itself: technology is getting away from us, so that *we are no
longer masters of our own mastery.* We have at our disposal, for example,
to mention a theme that preoccupies Jonas, the means to execute
genetic modifications capable of altering both animals and humans.
And there is no way—ethical, legal, political, or other—to control
the experiments we know are taking place everyday in laboratories
the world over. What is more, there is no reason to doubt that, due to
obvious economic incentives (considerable sums may ride on the discovery of a vaccine, of a medication, and so on), an increasing number
of disturbing experiments will continue to be performed without our
knowledge. Thus the creature escapes his master and may end up
irreversibly subjugating him. It is, therefore, necessary to establish a
*third power,* to again master the mastery of nature. But the task seems
impossible, or at least infeasible, according to Jonas, within the
framework of a democratic society. We must have recourse to force—
and here we return to the conclusions reached by *Greenpeace*—to State
constraint, for example, which Jonas cannot help but admire and
encourage in Asia and formerly in the Soviet Union . . . [23]

[23] See Hans Jonas, *The Imperative of Responsibility* (Chicago: University of
Chicago Press, 1984).

Indeed, the entire question lies herein. No serious democrat will argue with the idea that it is necessary, if not to limit the deployment of technology, at least to *control and direct it*. The idea that this control must occur at the price of democracy itself is an additional step which deep ecologists, propelled as they are by a hatred of humanism and of Western civilization, but also by a nostalgic fascination with models of the past or potential models of the future (the Indians, communism), almost never hesitate to take. There are some reasons which are beginning to appear for this tendency, I hope, and which a certain love of life, which goes hand in hand with that of nature, only reinforces.

### *"Biocentrism," or The Cult of Life*

The love of life—one's own life and that of those who are dear to us—is obviously one of the passions most common to humanity. There is nothing shocking about this, nothing that could pass for the bias of a particular ideological sensibility. But when this attachment is transformed into "vitalism," when it involves what deep ecologists call the "biosphere," its meaning shifts. The term "biosphere" warrants clarification. It does not designate, as one might think at first glance, the simple totality of living beings. Instead it defines the ensemble of elements within the ecosphere that contribute to the maintenance and blossoming of life in general. Why is this nuance essential? Because it allows one to differentiate between a love of life that is still "homocentric," reserved for human beings, and a "holistic" love, involving the "biogenic" whole upon which our existence directly or indirectly depends.

In plain language, it is a matter of recognizing, here again, that man is just an infinitesimal part of the universe, that he is wholly dependent upon it and that, given this, he must respect and value this universe more than he ordinarily does humanity. As our fundamentalists tirelessly insist, "Recognition of this fundamental dependence should raise the value of the life-sustaining milieu to that of an *end* . . . Some of the world's ecosystems should be strictly preserved, and their components should be given names other than 'resources' to indicate that they are sacrosanct." This would allow us to put an end to the "ecologically naive tradition of people valuing only people, supported by a homocentric ethic passed down from a long humanistic history in which environment-as-nature has been viewed with distrust as an alienated world, as less than human, as inferior and unworthy of fundamental valuation."[24]

I have often been struck in the course of my readings by the frequency with which religious expressions—"sacrosanct values," the "*Sanctity of Life*," and so on—pop up in the writings of deep ecologists when it comes to evoking the living in general. It should be clear that this fact can mainly be explained by the holistic nature of this body of thought: wishing to transcend the limits of humanism, it comes to consider the biosphere as a quasi-divine entity, infinitely more elevated than any individual reality, homan or nonhuman. Both exterior to men and superior to them, it can ultimately be regarded as their true creation principle—a notion that brings us back to one of the classical images of divinity. *Deus sive natura,* Spinoza's pantheism was already saying . . .

But this new Spinozaism also connects with one of the most profound intuitions of Nietzschian vitalism, according to which life constitutes "the most intimate essence of the being," the ultimate foundation of all things as well as the basis for all valorization. Recall that in the name of such a reference to life, Nietzsche came to denounce "the absurdity" of the Platonic-Christian opposition between this (tangible) world and the (intelligible) world beyond. According to him, this dualism merely conceals a pathological and "decadent" desire to negate real existence, which is nonetheless *the only life that truly is,* in favor of a pure fiction produced by the lucubrations of a sick imagination. Such is the essence of morality and religion, forever destined, neurotically so, Freud would later say, to seek a meaning to life *elsewhere:* "The resulting 'other world' is synonymous with non-being, non-living, with the will not to live. *On the whole,* it is *weariness with life,* and not the vital instinct, that created the 'other' world."[25]

This is why *there can be no value external to life,* a theme crucial to

[24] Rowe, "Crimes against the Ecosphere," p. 89.
[25] Friedrich Nietzsche, *The Will to Power,* vol. 1, book 1, (Paris: Gallimard).

deep ecology and which has also been adopted, as we shall later see, by "ecofeminism": "Judgements, judgements of life, concerning life, for it or against it, can, in the end, never be true: they have value only as symptoms, they are worthy of consideration only as symptoms; in themselves such judgements are stupidities."[26] For life is always expressing itself in us and through us, even in our most seemingly ethereal statements. Hence the idea, so dear to deep ecologists, that we must ultimately re-learn our reverence for the quasi-sacred entity which commands the alpha and omega of our existence, instead of rebelling against it with the vain and stupid temerity Western civilization has shown. One must accept the reality of our total *immanence* to nature, to this biosphere, revolt against which can only be pathological, thus provisional, and destined to fail.

Here again, deep ecology reveals its two faces. For this holistic-type vitalism, whose admitted target is humanism, lends itself to two political readings. One, more or less neoconservative and counterrevolutionary (hostile to the Declarations of Human Rights), can claim its roots in German romanticism, or even in the French extreme right. Monsieur Taine's tree, depicted in Barrés's *Les dérachinés,* might serve as an illustration for it: the plant, not only a "powerful mass of greenery" but a teacher of morality as well, is an object of admiration in that it "obeys a secret reason, the most sublime philosophy, which is the acceptance of the necessities of life." But the idea of taking the living being as one's model, or at least of imbuing it with the objectives of an authentic existence, now has its "leftist" version. Hans Jonas, for instance, whose professions of faith in favor of Communist regimes we have noted, has undertaken to establish that the goals of his ethics of responsibility "reside in nature." If ultimately one must protect the environment, preserve it for future generations, deep down it is because "life says yes to life," because there already exists, both within us and without us, in the animal and vegetable kingdom, a choice toward preservation of the being rather than suicide. One may object that Jonas's thought, and that of the German ecologists who claim allegiance with him, is deeper,

[26] "The Problem of Socrates," *Twilight of the Idols,* translated by Walter Kaufmann (New York: Viking Press, 1954), par. 2.

that it extends past the boundaries of a simple philosophy of biology. One may emphasize, for example, that it attributes to human freedom, and not just to living nature, the distinctly ethical task of *voluntarily* taking responsibility for preserving the world. This is to overlook the romantic, antihumanist underpinning that has been reinvested into this new philosophy of nature.[27] Jonas nonetheless explicitly mentions it over and over again: if it is true that the "yes" to life and the "no" to death which characterize all living beings only become conscious, and thus responsible and voluntary, in the context of humankind, it is nonetheless true that this desire and this conscience *belong to it only insofar as it is a living species culminating the trajectory of natural evolution, thus, as a reality of nature more than of freedom:* it is only "*as the supreme outcome of nature's purposive labor*" that "man must adopt the 'yes' into his will and impose the 'no' to not-being on his power."[28]

As in Nietzsche or Schelling then, the whole business of transcendence is done away with and our attitudes are ultimately just "symptoms," products of life in general. A simple but crucial question is thus raised: If there is nothing *beyond* life, why continue to admit the existence of values situated *above* it, of ideals in the name of which one might still dream of making what a vain and outdated morality until recently referred to as "the supreme sacrifice"? Once senses perhaps that the link between ecology and pacifism, between the concern for protecting life and that of not risking it, is deeper than it may have seemed at first glance.

[27] Historians of German idealism will have no trouble recognizing the influence of Schelling's theses on Hans Jonas.

[28] Jonas, *The Imperative of Responsibility,* p. 82. Paul Ricoeur, in an article on Jonas published in the *Méssager Européen,* no. 5 (1991), accurately perceives the difficulty: basing ethics in biology is insufficient, for the fact that nature "says yes to life" does not imply an *ethical* necessity that men act in favor of its preservation. It is necessary, according to Ricoeur, for Jonas to accept a humanist, Kantian-type moment. Unless, I would add, one can integrate freedom, conscience, and desire into nature itself, drawing inspiration, as Jonas deliberately does, from a Shelling-like romantic philosophy of nature, in which man appears as the summit of natural evolution more than as a being opposed to nature.

*Fear as a Political Passion*

At the base of contemporary ecology lies a "great planetary fear" which the authors of a recent work propose dividing into three categories: depletion of natural resources, growth of industrial, and particularly nuclear waste, and destruction of traditional cultures.[29] Factual and empirical fears, in short, relating to dangers whose exact extent and reality should be scientifically measurable: the greenhouse effect, holes in the ozone layer, possible explosions of radioactive cores, the disappearance of the Amazon forest and its inhabitants, the pollution of the seas, and so on. But there is more here than a *concern* for preserving nature as it is, or even of restoring it to the way it was before, in order to pass it along, if possible intact, to future generations: there is the founding principle of a political platform.

This at least is what Hans Jonas meant to describe in discussing what he calls a "heuristics of fear." The idea, less obvious than it seems at first glance, deserves special attention, as it crystallizes a major theme of fundamental ecology. What is it about? First of all, about becoming aware of the formidable gap between the weakness of our knowledge on the one hand, and the extraordinary potential for destruction at our disposal on the other. Jonas is not thinking so much about any specific danger here, about the nuclear threat, for example, as about the development of technology in general. This technology—like Frankenstein's monster or the creature who escapes from the sorcerer's apprentice—develops capacities to annihilate the earth which are especially terrifying since, as in the myths just mentioned, they emancipate themselves from any possible control by man. In the time of the Ancients, even up until the eighteenth century, a kind of harmony was preserved between our knowledge and our power: men lived in a world that was less complex than the one today, and their power over this world was infinitely smaller. The relationship has reversed itself: not only do we possess the means to liquidate all life, but the complexity of our universe is such that in most cases it is impossible to measure the consequences of our technological, economic and political decisions.

[29] Pierre Alphandéry, Pierre Bitoun, and Yves Dupont, *L'équivoque écologique,* (Paris: La Découverte, 1991), pp. 101 ff.

Hence the ethical and even theoretical function of fear, which becomes simultaneously a moral duty and a principle of knowledge. A moral duty, because we do not have the right, according to Jonas's thesis, to take any risk, however slight, that might be a *total* risk, by which he means a risk liable to endanger the very possibility of human existence and, more generally, of life; a principle of knowledge as well, since fear becomes our guide in detecting dangers of this type and allows us to distinguish them from those which are less absolute and whose risk might be acceptable. Thus the problematics of the concern for future generations is introduced into ecology. As Jonas suggests, fear is already inherent in the original question with which one may imagine that all active responsibility begins: what would happen to *him* if *I* don't take care of him? The more the answer is unclear, he maintains, the plainer the responsibility becomes.

This last concept recalls the central theme of his "heuristics": our responsibility is all the more urgent as our knowledge of the unexpected consequences (the "side effects") of our actions is quite small. This marks the difference with classical political philosophies, in particular with Hobbes, who also made fear a founding principles of politics. Indeed, one may recall that the state of nature is described, in Hobbes, as the place of the "war of all against all." This is to say that before the appearance of laws governing life in society, men lived in constant fear of a violent death. And it was to escape this feeling, to obtain security, that he accepted to enter into a State governed by laws. Fear is thus the basic political passion. Nevertheless, his motivation is still egoism. Jonas would like instead to convince us that in contemporary ecology, contrary to the scenario depicted by Hobbes, we are dealing with a "fear for others," notably for future generations.

It is of course impossible, and probably not very desirable, to sound out people's hearts and wills. This type of investigation always ends in putting intentions on trial. It seems, nonetheless, that Jonas's altruistic injunction is wishful thinking, and this, I believe, for one basic reason: it is difficult to see how the feeling of fear, precisely because it is a feeling, can be anything but primarily egocentric. Future generations all too often boil down to the image of our own children, and the concern with preserving life in general becomes confused with that of preserving one's own and that of one's family. Nothing could be more natural. But we should also think about this: ecology seems to be the first revolutionary political movement that bases itself on doing away with the risk of death and is in every way hostile to utopias (it is significant that Jonas's book is openly directed against a book by Ernst Bloch, whose title he even parodies). In every youth revolt, up to and including May, 1968, heroism was, one might say, a requirement—or at least, its display was a necessary rite of passage. Once life becomes the primary value before all others, once transcendency disappears because there is nothing beyond or below the biosphere, one can understand the preference for being "better red than dead." This, no doubt, is the price for ecologist pacifism in terms of ethics.

*Ethics and Science: The Return of "Objective Morals"*
Is there such a thing as an expert in matters of morality? The question would be amusing if not for the fact that it is becoming more and more real every day. In the United States, Canada, and Germany, it has become an "academic" topic, giving rise to university colloquia and publications. But in the rest of society as well, we are increasingly hearing the "advice of experts"; "ethics committees" are being created on which scientists, jurists, philosophers and theologians are called to give opinions on questions affecting people's private lives: medically assisted procreation, organ transfers, human experimentation, euthanasia, and so on. We are witnessing the development of the idea that knowledge of the secrets of the universe or of biological organisms endows those who possess it with a new form of wisdom, superior to that of mere mortals. But it is probably in the area of ecology that the feeling that the natural sciences will delivery *ready-made* teachings applicable to ethics and politics seems to be most confidently asserted. The problem of how to move from theory to practice is a classic philosophical one, but here it is finding new and current applications that merit real reflection. For there is always considerable danger that a new dogmatism will resurface when one claims to have found "natural," thus "objective," models of behavior, and to be able to decide *more geometrico* where good and evil lie.

The argument developed as early as the eighteenth century by the Scottish philosopher David Hume is nonetheless well known: from the simple consideration of what *is,* it is impossible to infer what *ought to be.* In plain language, a scientific theory may well describe reality to us as adequately as possible; it may plausibly anticipate the eventual consequences of our actions, still, in practice, we can draw no *direct* conclusions from it. Even if the surgeon general has convincingly determined that tobacco consumption is harmful to our health, an intermediary link is required before any ethical conclusions can be drawn: indeed, good physical health must be considered a *value* beforehand in order for the results of the scientific work to take the imperative form "don't smoke!" It is thus always *subjectivity* (an "I" or a "we") which ultimately decides whether to value a particular attitude or not. Without a *decision,* the imperatives one claims to draw from the sciences remain "hypothetical," since they cannot go beyond the framework of a formulation such as: "*If* you don't want to harm your health, *then* stop smoking." Because it is possible, after all, at least in a situation such as this in which one's personal well-being is concerned, to have values other than those of self-preservation, to prefer a life that is short but good, for example, to one that is long and boring.

If Hume's arguments are credible, we are forced to agree that morality cannot, *in and of itself,* be left to the experts. Certainly, those knowledgeable in certain matters may have a role to play in the determination of our choices when it is important to take into account the consequences of our actions and when these consequences are difficult to predict. To stick to the classic examples, a military chief who leads his troops to a certain death because he did not heed his well-informed advisors commits not only a strategical error but a moral one. A politician who, for lack of insight into economics, condemns a portion of the population to unemployment, would find himself in an analogous situation. Certain errors, those that could have been avoided by taking into account the available and accessible knowledge, can thus be regarded as faults. The borderline between acceptable ignorance and ignorance that may be deemed guilty is difficult to determine: hence the efforts of contemporary philosophy to reformulate the terms of an ethics of responsibility. Nonetheless it remains the case that once the importance of his input has been situated and accepted, *the expert as such is not the one who determines the choice of values.* This is a lesson we would be wise not to forget, for, from Lenin to Hitler, the notion that one is basing one's actions on an objective science of nature or of history has always ended in human catastrophe.

And yet this is the very gap radical ecology means to bridge, and to do so from at least three different perspectives.

First, from that of utilitarianism, which today is the main doctrine to revalorize the idea of a "moral expertise": if one accepts that interests can be calculated (the postulate, of course, is a bit difficult to accept, but without it everything collapses in this doctrine), then the "mathematician of passions" will be the moral expert par excellence. Thus, for example, utilitarians will discuss the comparative worth of the suffering of children, animals, or the mentally ill, in the hope that an exact science of pleasure and pain will finally allow us to make rational ethical choices.

Next, it is from the perspective of a philosophy of life—or of biology—that ecology hopes to find an "objective" foundation for ethics. Because life, according to Jonas's formula, "says yes to life." Nature *in itself* contains certain objectives, certain goals—the preservation instinct, for example, and the will to "persevere in being"—independent of our opinions and our subjective decrees: "to ground the 'good' or 'value' in being," writes Jonas, "is to bridge the alleged chasm between the is and ought."[30] Man is no longer, as in the context of republican humanism, an *autonomous* creature who sees himself as the author of norms and law, but rather, as the most elevated product of nature, he is the one who deciphers, safeguards, and takes responsibility for them. Here, ecology begins to think in terms of the Aristotelian *Cosmos,* of the world order uncovered by theoretical wisdom in which one could still detect an immanent justice and *read the law,* that is, the share and place falling to each of us. Plagiarizing Jonas even in his terminology, Michel Serres unhesitatingly responds as follows to the journalist who questions him about the basis of values:

> The basis that you are seeking for the values to determine . . . our decisions is simple: to act in such a way that

[30] Jonas, *The Imperative of Responsibility,* p. 79.

life remains possible, that generations can follow one another, that the human race can perpetuate itself . . . The
life of the entire species is in our hands; it is a basis as true
and faithful to things as that of the sciences themselves.
We are entering a period in which morality is becoming
objective.[31]

"True," "simple," "global," "objective" . . . the foundations of
ethics? Indisputable, then? What good news! But might this only be
the case in the minds of those who would like to put an end to the
uncertainty characteristic of all democratic questioning?

Paradoxically, a third group of ecologists is also turning to
Hume and the empiricist tradition. This is a paradox that warrants
attention indeed, since Hume seems to be the philosopher who most
eminently bars the passage from science to ethics, from Is to Ought.
In a noted article, one of Aldo Leopold's disciples, J. Baird Callicott,
himself a deep ecologist, poses the problem in exemplary fashion.
Instead of simply transgressing the Humian interdiction, as do Jonas
or Serres,[32] he chose to show that it is possible to pass from *is* to *ought*
while remaining faithful to the principles of empiricism. This is a decisive undertaking, then, if one accepts that the positive sciences
upon which ecology would like to base its moral vision of the world
are *empirical* sciences. It is a demonstration that is especially necessary
as this scientific basis for an environmental ethic can only function,
according to Callicott, within the framework of Humian thought.
Here is his reasoning:

*a)* At first glance Hume's position seems unfavorable to the project. Yet his *Treatise on Human Nature* repeatedly insists on the fact
that human nature being for the most part the same in each of us,

[31] Michel Serres, in *Le Monde,* 21 January, 1992.

[32] On this perspective, see Holmes Rolston III, "Is There an Ecological
Ethic?" *Ethics* 85 (1975). See also, by the same author, "Are Values in Nature
Subjective or Objective?" *Environmental Ethics* 3 (1981); Don E. Marietta Jr.,
"Interrelationship of Ecological Science and Environmental Ethics,"
*Environmental Ethics* 2 (1979); Tom Regan, "On the Connection between
Environmental Science and Environmental Ethics," *Environmental Ethics*
2 (1980).

moral variation, like variations in taste, are far less significant than is ordinarily stated in support of skeptical arguments. In truth, the very limited differences are exceptions to be interpreted either as deviations from the natural norm (evil is associated in this case with the pathological) or as a lack of cultivation of our nature, which naturally benefits from being developed (evil is thus one of the faces of "primitiveness").

*b)* It is thus in Hume himself that we find the missing link: mediation between is and ought can easily be accomplished through the universal nature of man. To take the spoken example to which Callicott refers, one would not say to one's teenage daughter, "You *ought not* to smoke cigarettes because cigarette smoking *is* deleterious to health," but "1) Cigarette smoking is deleterious to health; 2) Your health is something toward which as a matter of fact you have a positive attitude (as today we would say; a warm sentiment or passion, as Hume, more colorfully, would put it); 3) therefore, you ought not smoke cigarettes."

*c)* The syllogism can be extended to ecology as a whole: "1) the biological sciences including ecology have disclosed a) that organic nature is systematically integrated; b) that mankind is a non-privileged member of the organic continuum, and; c) that therefore environmental abuse threatens human life, health, and happiness. 2) We human beings share a common interest in human life, health and happiness. 3) Therefore, we ought not violate the integrity and stability of the natural environment . . ."

Callicott's reasoning is at least coherent . . . within the Humian framework, that is. Outside of this framework, it encounters at least two difficulties.

The first is that its moral imperative is never merely hypothetical. Indeed, as biological beings we all care about our health. But only, it must be specified, "to a certain extent." For health is not an absolute value for everyone or in all circumstances. Although "assertoric," which is to say based on an empirical fact, Callicott's imperative, therefore, remains *relative*. For it is become absolute, more than just factual conditions need to be added, for instance: "You ought not to smoke, not simply because, since you care about your health, you already essentially don't want to, but because you have a moral obligation to live as long as possible to raise your children, to help those less fortunate or to perform other tasks to help others."

The neo-Humian thesis, then, does not go beyond the level of what one is forced to call an "ethnology," not an ethics: it is not a question of moral norms one should endeavor to apply but only of a nonnormative analysis of what humans, de facto (according to Hume or Callicott), are supposed to love or abhor. The ethical criteria becomes identified with what empirical anthropology teaches us about human (or Humian . . . ) psychology. But the danger of an alleged scientific basis for ethics resurfaces: starting with the idea that, in principle, each individual possesses a "healthy and identical" human nature, we are gradually led to associate all supposedly deviant practices with pathology. In the extreme, evil is confused with abnormality: one has to be crazy to smoke, not to love nature as one *should,* and so on—which is why Callicott can say without hesitation that anyone who denies the validity of the ecologist syllogism ought to be advised to see a psychological counselor, thus linking up, no doubt involuntarily, with one of the worst aspects of Marxism, that of deducing ethics from science.

ﻉ

A strange ideal-type: coherent, but difficult, even impossible to classify, what with the persistent intermingling of, on the one side, the love of the native soil, nostalgia for lost purity, hatred of cosmopolitanism, modern rootlessness, and the universalism of the rights of man; but on the other, the dream of self-management, the myth of zero (or, as one now says, "sustainable") growth, the fight against capitalism, racism and neocolonialism, and in favor of local power, popular initiative referendums, and the right to be different . . . The common thread among these themes, which seem scattered in every direction, often on the verge of being irreconcilable, runs deep nonetheless. As soon as one grasps the principle behind it, the ideal-type attains the coherence (if not the truth, which is another matter) we may have been inclined to deny it: *it is that in all cases, the deep ecologist is guided by a hatred of modernity, by hostility toward the present.* As Bill Devall writes, in a passage that betrays the basis of his thought: modern civilization is that in which "the new is valued over the old and the present over future generations."[33] The ideal of deep ecology would be a world in which lost epochs and distant horizons take precedence over the present. It is not by chance, then, that it continually hesitates between conservative romantic themes and "progressive" anticapitalist ones. In both cases, the same obsession with putting an end to humanism is being asserted in at times schizophrenic fashion, to the point that one can say that some of deep ecology's roots lie in Nazism, while its branches extend far into the distant reaches of the cultural left.

[33] Devall, "The Deep Ecology Movement," p. 301.

# Nazi Ecology: The November 1933, July 1934, and June 1935 Legislations

*"Im neuen Reich darf es keine Tierquälerei mehr geben (in the new Reich cruelty toward animals should no longer exist)."* Excerpted from a speech by Adolph Hitler, these sympathetic words inspired the monumental law of 24 November 1933 providing for the protection of animals (*Tierschutzgesetz*). According to Giese and Kahler, the minister of the interior's two technical advisors who were responsible for drafting the legislative text, the goal was to at long last translate this message from the führer into concrete reality—a task that would have been impossible, it seems, before the coming to power of a National Socialist government. This at least is what they explain in a book they published in 1939 entitled *The German Law for the Protection of Animals*.[1] Assembled in some three hundred tightly printed pages are all the legal provisions relative to the new legislation, as well as an introduction explaining the "philosophical" and political grounds of a project whose breadth was indeed unequaled at the time. It would soon be complemented, on 3 July 1934, by a law limiting hunting (*Das Reichsjagdgesetz*), then, on 1 July 1935, by a landmark of modern ecology, the law for the protection of nature (*Reichsnaturschutzgesetz*). The three, ordered by Hitler, who made them his pet projects—though they also corresponded to the wishes of numerous and powerful ecologist associations of the period,[2]—bear the signatures of the

[1] Giese and Kahler, *Das deutsche Tierschutzrecht* (Berlin: Duncker and Humblot, 1939).

[2] In particular of the *Bund Deutscher Heimatschutz,* founded in 1904 by the biologist Ernst Rudorff, and the *Staatliche Stelle für Naturdenkmalpflege in Preussen,* created in Berlin in 1906. On these associations and, more generally, on the nature protection movements under the Nazi regime, one should

principal ministers concerned: Göring, Gürtner, Darré, Frick, and Rust, in addition to that of the chancellor.

Curiously, while these laws were the first in the world to reconcile a broad ecological plan with the concern for real political intervention, we find no trace of them in today's literature devoted to the environment (aside from a few allusions made by adversaries of the Greens and particularly vague as they rely on second-hand references). We nonetheless are dealing with a very elaborate series of texts that is fully representative of a neoconservative interpretation of what would later be called "deep ecology." An analysis of these texts is, therefore, in order.

First let us be clear about our objective. The troubling proximity between the love of one's native soil which motivates certain fundamentalist ecologists and the fascist-leaning themes of the 1930s has often been highlighted. In the preceding chapters we have had an opportunity to consider the ways in which these comparisons may *sometimes* have been appropriate. But we must also be wary of the kind of demagogy that invokes the horrors of Nazism to disqualify any ecological concerns a priori. The presence of an authentic interest in ecology at the heart of the National Socialist movement is not, in my view, in and of itself, a pertinent objection to a critical examination of contemporary ecology. On that basis, we would also have to denounce the construction of freeways—which, as we know, was one of the priorities of Hitler's regime—as fascist. Guilt by association, here as elsewhere, is inappropriate.

That said, these important legislations nonetheless lead us to reflect on the fact that an interest in nature, while it may not imply a

---

read the works of Walther Schoenichen. A committed National Socialist himself, holder of the Chair for the Protection of Nature at the University of Berlin, he was writing a series of works until the late 1950s on Germany's mission in the matter, including two essays on the contributions of Hitler's regime: *Naturschutz im dritten Reich* (Berlin, 1934) and *Naturchutz als Völkische und internationale Kulturaufgabe* (Iena, 1942), which no doubt constitutes one of the best commentaries one can read on the significance of Nazi ecology in the eyes of those who were involved in developing it. In it, notably, the legislations are situated within the intellectual history of German romanticism.

hatred of men ipso facto, does not exclude one either. Hitler's words upon inaugurating the *Tierschutzgesetz* give one pause after all. Before entering into the exceptional content of these laws, we must examine the disturbing nature of this alliance between an utterly sincere zoophilia (it was not limited to words but was borne out in law) and the most ruthless hatred of men history has ever known. The fact that we will not use this observation to hastily condemn all forms of ecology must not prevent us from considering its significance.

The love of nature, such as deep ecology invites us to experience it, is accompanied, both among "reactionaries" and "progressives," by a certain penchant for deploring everything in the culture that results from what I refer to as "separation" (but which can also be designated pejoratively, if one prefers, as "uprootedness"), and which the Enlightenment tradition has always seen as a sign of that which is properly human. All forms of thought that consider man a *transcendent* being, whether Judaism, post-Hegelian criticism,[3] or French republicanism, define him as the antinatural being par excellence. Given this, it is not surprising that the Nazi draws his gun to shoot the stateless person, the person who is not rooted in a community, when he hears the word culture. It is not surprising either that he would do so while preserving intact the love of the cat or dog who shares his life.

In this respect, the philosophical underpinnings of Nazi legislation often overlap with those developed by deep ecology, and this for a reason that cannot be underestimated: in both cases, we are dealing with a same *romantic and/or sentimental* representation of the relationship between nature and culture, combined with a shared revalorization of the *primitive* state against that of (alleged) civilization. As the biolo-

---

[3] The school of Marbourg, but also Husserl's phenomenology, could serve as references here. With the notion of "transcendency" or of "eksistence" as peculiar to the *Dasein,* Heidegger also fit in with this tradition, which is in fact why his adherence to Nazism, though deep and lasting, was only *partial* and never extended to the "biologist" and vitalist side of the ideology. That many of his disciples are now seeking to eradicate this thought on what is peculiar to man, on authenticity, by which Heidegger still belongs (somewhat) to the tradition of humanism, is a sign of the times that does not auger well.

gist Walther Schoenichen, one of the primary Nazi theoreticians for the protection of the environment, has continually insisted, the 1933–1935 laws are the culmination of romanticism, "the perfect portrayal of the popular-romantic idea" (*die Darstellung der völkisch-romantischen Idee*).[4] Significantly, despite his aversion to the United States, land of liberalism and plutocracy—an aversion we find today still intact among many young German ecologists—he recognizes a kinship between the love of the "Wilderness" and that of "*des Wilden*": in both cases, a certain desire to return to a lost natural virginity is expressed in words that, attesting to a common origin, designate the same "*primitive state*." And Schoenichen salutes the mid-nineteenth-century development of American national parks as a decisive event in the development of a proper relationship to nature. He emphasizes, in all seriousness, that the designation itself is felicitous, since it at least includes one word that tends in the right direction . . . [5]

## The Two Conceptions of Nature

We are not trying to lend credibility here to the opinion that says that Nazism is the pure and simple continuation of romanticism and, as Schoenichen claims, its adequate realization. No doubt it would be absurd to consider Hölderlin or Novalis the founding fathers of National Socialism, just as it would be to view Stalin as Marx's most faithful interpreter. And yet, the principle of the Nazi laws do indeed embody a theme that is central to romantic sentimentalism's struggle against the classicism of the Enlightenment: true nature, which must be protected at all costs against the misdeeds of culture, is not the nature that has been transformed by art, and thus *humanized,* but the raw, virgin nature that bears witness to the origins of time. It is impossible to understand Nazi ecology without understanding that it was inscribed in the context of an age-old debate on the status of nature as such. It is important to briefly mention what was primarily at stake, which is crucial in this context.

As early as the mid-seventeenth century, two antinomic por-

---

[4] Schoenichen, *Naturschutz als völkische und internationale Kulturaufgabe,* p. 45.

[5] Schoenichen, *Naturschutz im dritten Reich,* p. 46.

trayals of nature appeared in the course of a debate that opposed the aesthetic school of classicism and that of "sentimentalism."[6] Not only were the status of beauty and art at stake in this debate, but our philosophical and political attitudes with respect to civilization in general, insofar as the process of cultural development distances us irretrievably, it is argued, from the alleged authenticity of our lost origins. For the classical artists, whose chosen homeland was France, this distancing was positive. What is more, for them the idea of an original and authentic nature had, in truth, no meaning. Here is why: beginning with Cartesianism and its struggle against medieval animism, the idea took form that true nature is not the nature we perceive directly through our senses but the nature we grasp through an effort of the *intellect.* According to Descartes, it is through reason that we apprehend the essence of things. And what the French classics would call *"nature"* is precisely this essential reality, which is opposed to the appearances that are readily able to be perceived. Thus Molière, who wanted "to paint from nature" in his comedies, did not describe the daily lives of ordinary men but sketched the *idealized model* of *essential characters:* the miser, the misanthrope, the Don Juan, the hypochondriac, and so on.

The archetype of this "classical" and rationalist vision is of course the French-style garden, which is based entirely on the idea that, to arrive at nature's true essence or, rather, at "nature's nature," it is necessary to employ artifice, to "geometrize" it. For it is through mathematics, by use of the most abstract reasoning, that one grasps the truth of reality. As Catherine Kinzler writes: "[T]he French garden, crafted, pruned, designed, calculated, overly subtle, artificial, and forced is ultimately, if we want to get at the bottom of things, *more natural* than a wild forest . . . What is presented for aesthetic contemplation is a cultivated, controlled nature, pushed to the extreme, more real and more fragile at the same time because its essence is only reluctantly revealed."[7] In the eyes of the French classics, there-

---

[6] I've analyzed the terms of this conflict elsewhere, in my book *Homo Aestheticus* (Paris: Grasset, 1990); [Luc Ferry, *Homo Aestheticus: The Invention of Taste in the Democratic Age,* translated by Robert De Loaiza (Chicago: University of Chicago Press, 1993).]

fore, the English garden is not natural: in the best of cases, it has only the appearance of being so. It does not attain the truth of reality. What is worse, it can turn to affectation and mannerism, since it incarnates neither nature in its brute state nor its essential mathematical truth. As for wild landscapes, forests, oceans, and mountains, they can inspire only horror in men of taste: their chaotic disorder conceals reality. While the harmony of geometrical figures evoke the idea of a divine order, nature in its virgin state presents to the mind only pagan images, bordering on the diabolical. This in fact is why the Alps, now considered a prime tourist attraction, were perceived throughout the classical age as a mere hideous obstacle that was burdensome to cross.[8] The beautiful, from this perspective, could lie only in the *artificial* presentation of a truth of reason, not in the display of sentiments that the restoration of an original state allegedly concealed by civilization might inspire in us. Nature should be disciplined, polished, and cultivated, in short, when all is said and done, *humanized.*

It was against this classical vision of beauty that the aesthetics of sentiment revolted. Far from being mathematical, crafted, and human, here true nature is associated with *original authenticity,* the feeling for which we have lost, as Rousseau would have it, due to the culture of sciences and the arts. Thus what is natural is not at issue here, as it was among the classics, but rather what *is not yet denatured,* what is in its "primitive state." Forests, mountains, and oceans reassert their place against the artifice of geometry. But there is more: not only can nature not be humanized through civilization—it only gets lost—but men, despite their pretensions, are part and parcel of nature. They must, therefore, remain faithful to it. Hence the defense by Rousseau and the first romantics of those who are designated, significantly, as "naturals": the Caribbean natives as yet uncorrupted by the taste for luxury and artifice, as well as those "proud, pure-hearted mountain men," whose

[7] Catherine Kinzler, *Jean-Philippe Rameau: Splendeur et naufrage de l'esthétique du plaisir à l'âge classique* (Paris: Minerve, 1983).

[8] See Robert Legros's wonderful introduction to the travel journal of the young Hegel in the Alps (éditions Jérôme Millon, 1988). I am restating here one of its fundamental themes.

very isolation has protected them from evil.[9] The myth of the golden age and a paradise lost returns. Accompanied, as is wont to occur, by the inevitable discourse on the "fall," foreshadowing the anti-humanist theme of the "decline of the West."

It has often been noted how far this aesthetics of sentiment still lies from romanticism in its maturity. The latter even presents itself as a synthesis of the opposition between classicism and the sentimental. Nature is defined as "Life," as the "divine" union of body and soul, of sensibility and reason. Yet the separation between sentimentalism and romanticism is less distinct than is ordinarily asserted: even in their philosophy of history the romantics would preserve the idea of a lost golden age, as well as the notion that beauty has more to do with sentiment than with reason.

It was essentially these two themes that Nazi ecology would retain, opposing French, rationalist, humanist classicism, full of artifice, and the *"German"* image of an original nature—primitive, pure, virgin, authentic, and irrational, because accessible *only through the paths of sentiment.*[10] This original nature is even defined by its extrahuman character. It is *outside of and prior to man:* outside of his

[9]Robert Legros perfectly described the birth of this new sensibility, which breaks with French classicism: "This nature is the nature of origins. It is 'original' in the sense that it is not yet tamed, organized, disciplined, subjected. It is all purity, innocence, blossoming, enthusiasm, freshness, spontaneity . . . And the mountain offers us the image of this 'original' nature, both virgin and plentiful. The effervescence of the flowers and the overflowing of rivers, the play of waterfalls and the wild herbs, the purity of the air and the fresh feel of the forests, here is nature in its true state, a nature that is not yet denatured . . . It is not only manifested in the Alpine landscape but also in the customs of the mountain folk. Living in harmony with original nature, the inhabitants of the Alps are themselves impregnated with a 'natural' spirit, meaning that they are not corrupted by civilization, deformed by artificiality . . . Through the ideal of an originally pure and generous nature the myth of a golden age takes form in the heart of the mountains," ibid., p. 20.

[10]Alfred Bäumler devoted a chapter to the German specificity of the esthetics of sentiment as opposed to the French character of classicism in his work on *Das Irrationalitätsproblem in der Logik und Aesthetik des achtzehnten Jahrhunderts,* republished in Darmstadt by the *Wissenchaftliche Buchgesellschaft.*

mathematical reasoning, and prior to the appearance of artificial culture, for which human folly and vanity alone are responsible.

In his 1942 work on the *Protection of Nature as a Popular (völkisch) and International Cultural Task,* Walther Schoenichen specifies the appropriate way to understand the notion of nature from a National Socialist perspective. The information he provides is quite interesting: starting with the basic idea that it is "obvious" that "respect for the creations of nature is etched in the blood of Northern peoples," he deplores the fact that the word "nature" is derived etymologically from the Latin *"natura."* This origin is troublesome—too southern, almost French; Schoenichen prefers for it to be immediately replaced with the Greek *phuo,* which means "to grow, to be born," and which gives the noun *phusis,* from which the word "physique" is derived. This philological operation brings us to the following conclusion: "We can, according to the preceding, consider it certain that the concept of nature designates, first and foremost, *objects and phenomena which occurred on their own, without man's intervention."* Here we are at the opposite extreme from the "humanized" nature of the classics. And this is precisely the point for Schoenichen, who insists on the value and significance of the Greek etymology, according to which "the lack, even the exclusion, of any intervention by man is the trait that characterizes nature." Thus we can and must Germanize (*verdeutschen*) the word nature, changing it to *Urlandschaft,* "earth" or "original land"!

With such a definition, Nazi ecology essentially preestablishes a link between the aesthetics of sentiment and what would later become the central theme of deep ecology: the idea that the natural world is *worthy of respect in and of itself,* independent of all human considerations. Thus Schoenichen particularly emphasizes the texts of Wilhelm Heinrich Riehl, which foreshadow the "environmentalist" critique of utilitarian—hence *anthropocentric*—justifications for ecology: "The German people need forests. And even when we no longer need wood to warm the outer man . . . it will be all the more necessary to warm the inner man. We must protect the forest, not only so that the stove will not go cold in winter, but so that the pulse of the people may continue to beat in joyous, vital warmth, so that Germany will remain German." In good logical order, this deconstruction of the primacy of individual interests concludes with a clear and distinct call for the rights of trees and rocks: "For centuries, we have been bombarded with the idea that it is progress to defend the rights of cultivated lands. But now we are saying that it is progress to demand the rights of the wild nature next to these lands. And not only the rights of the wooded lands, but also of the sand dunes, swamps, garigues, reefs, and glaciers!"

## The Critique of Anthropocentrism and
## the Call for the Rights of Nature

Both of these elements are especially present in the most important law, which affects the protection of the animal kingdom, this "living soul of the land" (*die lebendige Seele der Landschaft*) in Göring's words. In it we find a long and meticulous analysis by the principal drafters, Giese and Kahler, of the radical innovations of the National Socialist *Terschutzgesetz* with respect to all prior legislation on this issue, domestic or foreign. By their own admission, this originality lies in the fact that, for the first time in history, the animal, as a natural being, is protected *in its own right, and not with respect to men.* A long humanist, even humanitarian, tradition defended the idea that it was indeed necessary to prohibit cruelty toward animals, but more because it translated a bad disposition of human nature, or even risked inciting humans to perform violent acts, than because it was prejudicial to the interests of the animals themselves. It was in this spirit that the Grammont law had prohibited the *public* spectacle of cruelty toward *domestic* animals (bullfighting, cockfighting, and so on) in France since the mid-nineteenth century.

If one compares the *Tierschutzgesetz* with the laws that were adopted in other European countries at the end of the 1920s, it does indeed seem to stand out in its clear desire to put an end to anthropocentrism.[11] It is necessary to quote the texts, which are exemplary in their precision:

> . . . the German people have always had a great love
> for animals and have always been conscious of our strong

[11] Only the Belgian law of 22 March 1929 can be compared to it, but England itself, to say nothing of the countries of southern Europe, does not punish cruelty toward wild animals.

ethical obligations toward them. And yet, only thanks to the National Socialist Leadership has the widely shared wish for an improvement in the legal provisions affecting the protection of animals, the wish for the establishment of a specific law that *would recognize the right which animals inherently possess to be protected in and of themselves (um ihrer selbst willen), been achieved in reality.*

Two indicators, which predominate in the inspiration for this new legislation, manifest its nonanthropocentrist nature. According to the drafters of the law (and aside from certain exceptions, including that of Belgium, they are correct in this), in all other legislations, including those in Germany before National Socialism, cruelty toward animals had to be performed *in public and against domestic animals* in order to be punished. As a result, the legal texts did not threaten "punishment serving to protect animals themselves, intending to defend them against acts of cruelty and bad treatment," but were meant in reality "to protect human sensibility from the painful experience of having to participate in an act of cruelty toward animals." It is now a matter of curbing "cruelty in and of itself, and not because of its indirect effects on man's sensibility." The lawmaker emphasizes that *"cruelty is no longer punished with the idea that one must protect men's sensibility from the spectacle of cruelty toward animals, men's interests are no longer the backdrop here, but rather it is recognized that the animal must be protected in and of itself (wegen seiner selbst)."* Acts of cruelty committed in private will, therefore, be just as reprehensible as others.

In the same spirit, Nazi legislation innovatively anticipates the most radical demands of contemporary antispeciesism by surpassing the anthropocentrically inspired opposition between domestic and wild animals.[12] This is the object of paragraph 1 of the law, which *"is valid for all animals. In the present law, the term 'animal' will be understood*

[12] We should note, nonetheless, that the drafters of the law refused to consider animals as legal subjects of the same rank as German citizens. But it is significant that the question is explicitly mentioned and discussed, and that the negative outcome does not result in the idea that animals do not have rights *in and of themselves.*

to mean all living beings designated as such by current language as well as by the natural sciences. From a penal standpoint, therefore, no distinction will be made between domestic and other types of animals, or between inferior and superior animals, or between animals that are useful or harmful to man." Thus we have a text which, at the opposite extreme from the Grammont law, could be signed wholeheartedly by today's deep ecologists.

Without entering into the details of this law, it should be noted that it carefully examines all the decisive questions that are currently being discussed by the defenders of animal rights: from the prohibition against the force feeding of geese, to that of vivisection without anesthesia, it proves to be fifty (or more) years "ahead" of its time.

Two further points on which the *Tierschutzgesetz* is particularly prolix and detailed again allow one to think that the love of animals does not imply that of men: one is an entire chapter devoted to the Jewish barbarity involved in ritual slaughter, which is henceforth prohibited. Another devotes inspired pages to the feeding, resting, aeration conditions, and so on, with which it is appropriate, *thanks to the blessings of the national revolution currently underway,* to arrange for the transportation of animals by train . . .

## The Hatred of Liberalism: Paradise Lost and the Decline of the West

The theme of the "fall," of "dereliction," is omnipresent in these laws. Original and authentic nature is contrasted with the destructive barbarity inherent to the modern liberal economy. This is emphasized from the start, in terms that are significant, in the preamble to the *Reichsnaturschutzgesetz* of 26 June 1935, thus reestablishing a connection with the romantic vision of a history in three stages—the golden age, the fall, the restoration:

> Today as before, nature, in the forests and the fields, is an object of longing (*sehnsucht*), joy and the means of regeneration for the German people.
> Our native countryside (*heimatliche Landschaft*) has been profoundly modified with respect to its original state, its flora has been altered in many ways by the agricultural and foresting industries as well as by the unilateral reallocation

of land and a monoculture of conifers. While its natural
habitat has been diminishing, a varied fauna that brought
vitality to the forests and the fields has been dwindling.

This evolution was often due to economic necessity.
Today, a clear awareness has emerged as to the intellectual,
but also economic, damages of such an upheaval of the
German countryside.

Before, one could only grant half measures to the
"natural monument" protection sites created at the turn
of this century because the people's vision of the world
(*weltanchauliche Voraussetzungen*) was wanting. Only the
metamorphosis of the German man was to create the
preconditions for a efficient protection of nature.

The German government of the *Reich* considers it
its duty to guarantee our fellow citizens, even the poorest
among them, their share in the natural German beauty.
It has, therefore, enacted the law of the *Reich* with a view
toward protecting nature . . .

A great deal could be said about this text. First, we find the
romantic confusion between the cultural and the natural that enables
us to make sense of ideas such as that of a "natural German beauty," or
of the "natural monuments" (*Naturdenkmale*), which paragraph 3 of
the law defines in terms reminiscent of deep ecology's plan to turn
wild areas into legal subjects: "Natural monuments, in the sense intended by this law, are original creations of nature, the preservation of
which is in the public interest due to their importance and to their
scientific, historical, patriotic, folkloric, or other significance—these
include, for example, boulders, waterfalls, geological accidents, rare
trees . . ." The law thus provides for the creation of "protected
natural zones" (par. 4).

But contrary to a persistent legend, what we see especially is that
while the Nazi regime was oriented toward modern technology, it was
also equally hostile to what we would now call economic "moderniza-
tion," perceived as destructive to particular ethnic qualities as well as
to original nature. It is here that we find a true "homage to difference,"
a reinstating of diversity in opposition to the one-dimensionality

of the liberal world. For the ideology that underlies liberalism, Schoenichen reminds us in the context of his defense and illustration of the 1935 law, is characterized by "the leveling influence of the general culture and of urbanization, which constantly and increasingly presses back the particular and original essence of the nation, while economic rationalization gradually eats away at the original specificity of the landscape."[13] According to a theme that would be readopted as much by German neoconservative revolutionaries as by sixties leftists, by Heidegger and by Marcuse, by Alain de Benoist and by Felix Guattari, it is necessary to learn to reindividualize, to redifferentiate groups and individuals, countering the vast tendency toward homogenization ("Americanization") which is the central dynamic of global capitalism. In its National Socialist version, this antiliberal theme translates into the idea that after the first two stages of history—the golden age and the fall—only the remaking of the German people (*die Umgestaltung des deutschen Menschen*) will lead us to the end of history, in other words, to the redemption that will enable us to rediscover our lost origins. Paradoxical as it may seem today, it is thus perfectly logical that the laws protecting nature would extend into a third worldism concerned with respecting the plurality (the "wealth and diversity") of ethnic differences.

## Third-Worldism and the Praise of Difference

We have to be ignorant or prejudiced not to see it: Nazism contains within it, for reasons that are in no way accidental, the beginnings of an authentic concern for preserving "natural," which is to say, here again, "original" peoples. In the chapter devoted to this subject in his book, Walther Schoenichen cannot find words harsh enough to condemn the attitude of "the white man, the great destroyer of creation": in the paradise he himself is responsible for losing, he has paved only a path of "epidemics, thievery, fires, blood and tears!"[14] "Indeed, the enslavement of primitive peoples in the 'cultural' history of the white race constitutes one of its most shameful chapters, which is not only streaked with rivers of blood, but of cruelty and torture of the worst

[13] Schoenichen, p. 21.
[14] Ibid., p. 411.

kind. And its final pages were not written in the distant past, but at the beginning of the twentieth century." Schoenichen proceeds to trace, with great precision, the list of the various genocides that have occurred throughout the history of colonialization, from the massacre of the South American Indians to that of the Sioux—who "were pushed back in unthinkable conditions of cruelty and infamy"—and the South African bushmen. The case of the latter is particularly symbolic of the misdeeds of liberal capitalism: they were killed because they had no notion of ownership. Game having disappeared from their region, this hunting people was forced to "steal" goats belonging to the colonists—the word "steal" must be placed in quotes, since bushmen had no concept of private property. And as they were thrown into prison without any idea of what was happening to them, they allowed themselves to die of starvation: "Thus an interesting people was exterminated before our very eyes, simply because an exogenous policy imposed on the indigenous population refused to understand that these men could not abandon their hunting lives to become farmers from one day to the next . . ."

This indictment, written in 1942 by a Nazi biologist who saw the *Naturschutzgesetz* as a means to remedy these misdeeds (does it not protect all forms of wild life?), is not without interest. Its designated target is liberalism and, more specifically, French-style republicanism. But it also has a positive goal: to defend the rights of nature in all its forms, human and nonhuman, so long as they are representative of an *original state (Ursprünglichkeit)*. On the first point, Schoenichen's attacks are clear. They throw in question capitalism's greed. For in the context of a different world vision, it "would have been entirely possible to find a reasonable compromise between the claims of the conquerors and the basic needs of the primitive peoples. It is primarily the liberal vision of the world that is responsible for having stood in the way of such a solution. For it recognizes no motivation other than economic profitability, which raises to the level of a moral principle the exploitation of the colonies for the sole benefit of the mother country." This naturally provides an occasion to assail the French theory of assimilation, which is, according to Schoenichen, "drawn directly from the principles of the

Declaration of the Rights of Man of 1789." Thus "the old liberal theory of exploitation always constituted the backdrop for French colonial policy, so that there was no room for a treatment of primitive peoples that tended in the direction of the protection of nature."

In opposition to this "assimilationist" vision of the primitive state, Nazi policy advocates an authentic recognition of differences: "The natural policy for National Socialism to follow is clear. The policy of repression and extermination, the models for which are furnished by the early days of America or Australia, are just as unthinkable as the French theory of assimilation. Rather it is appropriate for the natives to flourish in conformity with their own racial stock." It is necessary then, in all cases, to leave the natives to their own development. The only recommendation, which according to Schoenichen is obvious "from the point of view of a National Socialist vision of the world" is the prohibition against mixed marriages, precisely because they imply the disappearance of differences and the uniformization of the human race. Now as before, the extreme Right assails inbreeding in all of its forms, assigning to ecology the task of "defending identity," which is to say "preserving the ethnic, cultural and natural milieux" of peoples—beginning, of course, with one's own: "Why fight for the preservation of animal species while accepting the disappearance of human races through widespread inbreeding?"[15] Indeed . . .

Like the aesthetics of sentiment and deep ecology, which also place new value on primitive peoples, mountain folk, or Amerindians, the National Socialist conception of ecology encompasses the notion that the *Naturvölker,* the "natural peoples," achieve a perfect harmony between their surroundings and their customs. This is even the most certain sign of the superiority of their ways over the liberal world of uprootedness and perpetual mobility. Their culture, similar to animal ways of life, is a prolongation of nature; it is this ideal conciliation that the modernity issued from the French Revolution has destroyed and which it is now a matter of restoring.

[15] Bruno Mégret, lecture at a colloquia on ecology organized by the National Front on 2 November 1991.

## Of Nature as a Cultural Trait and of Culture as a Natural Trait

Thus the unity of nature and culture in the life of the German nation must be reinstated, each element passing into its opposite to find its true essence, in keeping with a romantic theme that refuses to separate the cultural and the natural, which was the tendency of Enlightenment thought. The authors of the 3 July 1934 law on hunting specify this in their introduction: "The two-centuries-old development of the German hunting law has reached an outcome of great importance for the German people and *Reich*. This legislation has not only enabled us to overcome the splintered state of the German law, which was reflected until now in twenty different regional laws, and to thereby arrive at a legal unity, but set itself the task of preserving wild animals *(des Wildes)*, *one of our most precious cultural assets,* and of educating the people to develop a love and understanding of nature and its creatures." Thus wild animals *(das Wilde)* are defined as one of Germany's "cultural assets," not as something that preceded all civilization. Conversely, the love of nature, a cultural trait if ever there was one, is presented as being rooted since time immemorial in the biological constitution particular to Germanity:

> The love of nature and its creatures as well as the pleasure of hunting in the forests and fields is deeply rooted in the German people. This is how the noble art of German venery developed over the centuries, backed by a Germanic tradition that has existed since time immemorial. Hunting and game must forever be preserved for the German people, as they are among our most precious assets. We must deepen the German's love for our national territory, reinforce his vital energy and bring him rest after the day's work.

Fishing, hunting, and tradition . . . We should specify right away that the goal of the law is not only to introduce legal unity to culture and nature, but also to situate it within the framework of an authentic ecological system of thought. Thus it is necessary to *limit* the right to hunting in order that it concur with the stated need for preserving the natural environment. In this sense, the law of 1934 is undoubtedly the first to redefine the role of the hunter in modern terms. According to a theme destined to a long posterity, the hunter goes from being a simple predator to one of the main architects of environmental protection, even of a restoration of original diversity, forever threatened by modern uniformization:

> The duty of a hunter worthy of this name is not only
> to hunt game, but also to maintain it, to care for it so that
> a healthier state of game, stronger and more diversified in
> terms of species, will develop and be preserved.

The sixth section of the law is devoted to establishing limitations on the right to hunt, limitations based not only on the need for security, public order, and even the protection of the landscape, but also on the need to "avoid cruelty to animals." It is in the name of this desire, dear to Hitler himself, that certain types of hunting using painful traps were prohibited. The *Reichsjagdgesetz* turns out to be the key pin of the National Socialist ecologist platform: in it, man is no longer positioned as master and possessor of a nature which he humanizes and cultivates, but as *responsible* for an original wild state endowed with intrinsic rights, the richness and diversity of which it is his responsibility to preserve forever.

## In Praise of Difference, or
## The Incarnations of Leftism:
## The Case of Ecofeminism

The love of diversity, the concern with preserving the variety of natural species and traditional cultures for their intrinsic value are now
no longer the prerogatives of a romantic extreme Right. The rejection of uniformity characteristic of modern times, the disdain for
mass consumption, the critique of cosmopolitan universalism have
even become more "leftist" concerns. Previously, the call for a right to
difference fell squarely within the realm of the counterrevolution:
faced with the formal abstraction of the "Declaration of the Rights of
Man," the first "nationalists" demanded a return to the English,
French, or Italian legal traditions . . . in short, to that of each *concrete*
and *distinct* community. How did this attachment to one's unique
heritage, this love of "home" which plainly placed itself in the camp
of the "reaction," become classified under the banner of "progres-
sivism?"

The answer is clear and has more to do with the weight of history than with the logic of ideas. It can be expressed in one word: de-
colonization. Was the conquest of the peoples of the Third World
inscribed in the logic of republican universalism? Or was it, on the
contrary, a betrayal of its most noble principles? All things considered, the answer matters little in terms of history, even if it remains
critical on a philosophical level. Whatever one may think of it, in effect, the opinion has taken hold that the republican idea concealed
the worst: an imperious will to assimilate, even eradicate, differences.
One need only travel in the countries of North Africa today to learn
that the slightest allusion to the secular and democratic ideal such as
it was developed in Europe risks seeming the sign of a nasty ethno-

centrism. We have entered into an era of widespread suspicion regarding the "Enlightenment"—for the better, sometimes, but often for the worse. Simultaneously, the ideology of the right to be different has ceased to be or to seem "reactionary" and has instead become associated with the entirely natural aspiration of the colonized to reclaim an individuality flouted by Europocentrist imperialism.

Thus the desire to restore the community identities of nations bound by a "popular spirit" and hostile to the geometrical tendency of the Jacobins is not longer an expression of an Ancien Régime–type nostalgia. On the contrary, it is the voice of the oppressed, of the "damned of the earth," demanding access to a culture the colonizer did everything in his power to obliterate. To teach about "our ancestors the Gauls" no longer goes over well in the classrooms of the former French colonies. . . . Hence the incredible paradox which is far from over: the closing in on oneself, the most extreme communalism and nationalism, have taken the form of a revolution! And the process, which cut its teeth in the Third World, is now being employed by certain Eastern republics in their struggle against the remains of the former Soviet empire. Everywhere the same cry is sounding: be good Muslims, good Czechs, and good Slavs, but also good Frenchmen, and, as in 1914, good Germans . . . And no one can say with any certainty whether the exhortation is "Right Wing" or "Left." Only history and sociology allow us to distinguish between the positions of the formerly oppressed and the new fascists, which tend to overlap. The worst evil is supposedly inbreeding and the bogus universalism of the worlds of technology, culture, and mass consumption. One watchword serves both the extreme Right and Left: let us learn to "reindividualize" our ways of life, to redifferentiate them in protest against the "one-dimensionality" of the modern world.

As the case may be, and even if we track the way in which ideas pass from one extreme to the other, we have to admit that ultimately, differencialist arguments tend to resemble one another. The love of the environment provides shining proof of this today, for the praise of difference, omnipresent in Nazi legislation, reappears almost word for word in the most "advanced" versions of deep ecology. Let us be clear: making such a connection does not imply that leftism and fascism are similar ideologies. This would be an insult, if one may say so, to both. It is absurd and simply unfair to compare the Green party to Vichy, as is often done in the polemical spirit of the day. We should beware the type of insult from which the object emerges unscathed, which, in its excess, loses its effect.

The fact remains nonetheless that the project of preserving certain communal identities, which is legitimate in and of itself, can sometimes have consequences that are all the more worrisome as they are not accidental, but inscribed within the very heart of a philosophy of difference. This is demonstrated by one text, among many others, written in December 1985 by Jean Reynaud and European Green deputy Gérard Monnier-Besombes, for the opening of the party's debate on immigration. In this context, it warrants our full attention:

> There is some *incoherence* on the part of ecologists in demanding the safeguard and promotion of regional cultures, without taking into consideration the strong presence of foreign minorities.

> It would be frustrating to visit Brittany if with every step one had to bump into a Swiss or a Croatian, if one could no longer find, even interspersed among other elements, a people (*ein Volk!*) in intense symbiosis with the land (*ein Land!*) it inhabits and to which it is attached, the linguistic, cultural and architectural heritage of which it can bring to life.[1]

Leaving aside the references to Nazi jargon, which are not in the best taste, the important thing in this declaration, what makes it feel like a true confession, is that its authors themselves insist on the *incoherence* of defending a differencialist position, yet without drawing the

---

[1] The complete text of this interesting article was given to me by J. F. Bizot, director of the magazine *Actuel,* on the occasion of the suit they brought (and lost!) against certain members of the Greens, who were scandalized at having been publicly called fascists.

logical consequences as to immigration. For from the point of view of a call for purity, if not racial at least cultural, the excessive mixture of Bretons, Croatians, and Swiss would appear to be intolerable!

This doctrine is clarified in another "document" which, though written by Antoine Waechter, could easily have been signed by a Walther Schoenichen. Let the reader be the judge:

> Civilizations are diverse, in the image of the living territories in which they are rooted. Like biological diversity, this cultural diversity is threatened because certain civilizations, convinced of their own superiority, aim to occupy all the land, declaring their values universal . . . Attachment to a community identified by its way of speaking, its traditions, its *savoir-faire,* its history, and the love of a territory which through its landscapes expresses the soul of this community, is one of the fundamental human dimensions. Being uprooted is a trauma, a source of psychological destabilization and existential difficulties. The first fundamental right of an individual is to possess an identity, and this identity is intertwined with that of the human group to which he belongs.[2]

Here again, let us avoid false accusations: no one can seriously suspect Waechter of sympathizing with fascism. The question, therefore, becomes all the more pressing: Why, then, such proximity with some of its most fundamental themes? Why speak in this way of "the soul of a community," compare culture with life, deny the primacy of individual autonomy in favor of community membership, sing the praises of roots and the land against the abstract universalism of this "Jules Ferry Republic which forbids the use of maternal languages and the display of distinctive signs of regional identity . . ."? Why, further, this marked concession to the ideology of "blood and soil": "A cultural community can only blossom on a soil where a continuity of generations occurs and here its identity takes the concrete and visible form of a unique landscape." What about those who are "stateless" or "uprooted," who do not correspond to these requirements?

[2] Waechter, *Dessine-moi une planète,* p. 161.

What about all the ways in which modern culture refuses to *root itself,* to base itself on the certainties of heritage or to identify itself with any particular region? Why praise difference when it takes the form of such blatantly nationalist or communalist claims?

There is no use accusing me of exaggeration here, sniffing out disingenuous and hyperbolic rhetoric, looking for quotes taken out of context and cleverly strung together. For "the praise of diversity," to borrow the title of a chapter in Waechter's latest book, is at the heart of fundamental ecology's intellectual purview. It can be found in Bill Devall,[3] Hans Jonas, Aldo Leopold, or Felix Guattari. One may of course object that there is nothing shocking about such praise in and of itself, that one-dimensionality and uniformity are indeed real dangers of the world of technology, especially in a country of Jacobin tradition such as France. Indeed . . . Here a minimum of philosophical clarification is in order. Because the respect for difference can be conceived in various fashions. And while no one wants to see a Europe in which the diversity of national cultures permanently vanishes in favor of an "American-style" standardization of lifestyles, still it is not clear that the solution is in the valorization of "dissensus" as such—especially one that seeks its origins in a community-based vision of relations among men.

Yet this is the thesis proposed to us by an "ultraleftist" version of radical ecology. We find it best formulated in the writings Felix Guattari devoted to ecology after joining Antoine Waechter's party in 1985. In an article written in collaboration with Danny Cohn-Bendit, he articulates his program in a manner that at least has the merit of being clear:

> The goal is not to arrive at a rough consensus on a few general statements covering the ensemble of current problems, but, on the contrary, to favor what we call a culture of dissensus that strives for a deepening of individual positions and a resingularization of individuals and human groups. What folly to claim that everyone—immigrants, feminists, rockers, regionalists, pacifists, ecologists, and hackers—

[3] "Diversity is inherently desirable both culturally and as a principle of health and stability of ecosystems" (Devall, p. 312).

should agree on a same vision of things! We should not be aiming for a programmatic agreement that erases their differences.[4]

Dare I admit it? I do not see what is so foolish about seeking agreement, nor how such a goal is ineluctably destructive of differences. On the contrary, it seems to me that according to the perspective taken by Guattari, these are all values of the *res publica,* the public space at the heart of which alone it is possible to freely construct, through discussion and argumentation, the *consensus* of law and of *common interest,* which are being entirely brushed aside in favor of a discourse that could easily belong to the new Right. For if different human groups and cultures neither can *nor should* seek to communicate with one another, if all reference to common values is tyranny and the destructive influence of universality, few options remain: we will witness, as in certain countries of Eastern Europe, the dissolution of the very idea of a republic in favor of a return to the romantic vision of communities viscerally closed onto themselves, incapable of surpassing their atavistic individualities to enter into communication with others.[5]

Under these conditions, it is no accident that the praise of difference, understood in this anti-Republican fashion, works its way into formulas which are, if not racist, at least racialist. Let's listen to Guattari again:

[4] Felix Guattari and Danny Cohn-Bendit, "Contribution por le mouvement," in *Autogestion, l'Alternative* (weekly publication of the PSU) (November 1986).

[5] Hence a veritable aversion to "discussion" which is expressed by G. Deleuze and F. Guattari in their latest book, *Qu'est-ce que la philosophie?* (Paris: Editions de Minuit, 1991). In it we read that theoreticians of the ethics of communication (this refers primarily to Jürgen Habermas, but not only to him . . . ) tire themselves out "searching for a universal liberal opinion as a consensus, beneath which one finds the cynical perceptions and affectations of capitalism"! Habermas would appreciate that . . . In the interview granted the *Nouvel Observateur* upon the publication of the book (yes, one must sacrifice to the rules of the game and give in to the imperatives of "communication" after all), we find this statement as well, which gives one pause: "To discuss is a narcissistic exercise in which each one struts his feathers in turn: soon you don't know what you're talking about

The various practices not only do not need to be homogenized, brought into line with one another under a transcendent tutelage, but should instead be engaged in a process of *heterogenesis*. Feminists can never do enough for women's advancement, and there is no reason to ask immigrants to give up *their national affiliation or the cultural traits that cling to their being*.[6]

"Cultural traits that cling to their being": Is this not a "leftist" version of racism? For Guattari as for Schoenichen, culture is an ontological reality, not an abstraction: it is inscribed in men's beings, just like their biological state—which is why it is necessary above all else to renounce the republican project of integration. It is in this way, once again, that the ideology of the right to be different forms a bridge between ecology and feminism—or at least between a certain ecology and a certain feminism. For another tradition of thought, that of existentialism, offered us a diametrically opposed fashion of conceiving of women's dignity. It is necessary to restate the principle here if we wish to fully perceive the stakes of a debate which can justifiably be considered as exemplary of a differentialist approach toward nature and of an ultraleftist version of deep ecology.

In *L'existentialisme est un humanisme*, Sartre devoted several pages of shining simplicity and depth to the classical question of "human nature." Without knowing it—he was hardly inclined toward the history of philosophy—he rejoins the fundamental intuitions of Rousseau and of Kant concerning the difference between the animal kingdom and humanity: man is the antinatural being par excellence, the only one capable of not being fully determined by the natural

---

any more. The difficulty lies in determining the problem to which a given proposition corresponds. Now, if you understand the problem posed by someone else, you have no desire to have a discussion with him: either you pose the same problem or you pose another one and would rather move in your own direction. How can you have a discussion if there is no common set of problems and why have a discussion if there is?" Indeed . . .

[6] F. Guattari, *Les trois écologies* (Paris: Galilée, 1989), p. 46.

conditions allotted him at birth. This is why, according to Sartre, there is no "human nature" strictly speaking. Let us avoid from the outset a misunderstanding that is only too frequent: it is not a matter of denying the incontestable fact that we live in a body, male or female, in a particular nation, culture, and social milieu, all of which have considerable influence over us. It is clear that I can no more choose my sexuality than I can my mother tongue. I receive them, so to speak, from without, and everything, clearly, does not result from autonomy: it would be simply absurd to assign freedom the impossible task of eradicating such external facts.

It is nonetheless true that we can, and what is more, we must, if we wish to be rigorous, create a distinction between a simple factual *situation,* even if it is intangible like belonging to one of the two sexes, and a *determination* which in some sense shapes us outside of all voluntary activity. For unlike an animal, which is subject to the natural code of instinct particular to its species more than to its individuality, human beings have the possibility of emancipating themselves, even of revolting against their own nature. It is by so doing, that is, by breaking away from the order of things, that one gives proof of an authentic humanity and simultaneously accesses the realms of ethics and culture. Indeed, both of these result from this antinatural effort to construct a properly human universe. Nothing could be less natural than rules of law just as nothing could be less natural than the history of civilization: both are unknown among plants and animals. One may say, along with Spinoza, that it is in the nature of big fish to eat the smaller fish. It is certainly to be feared that this may at times, even often, be the case, *mutatis mutandis,* among human beings, and that they may be inclined, *by nature,* to give in to force. But this is not what we expect of them on an ethical and cultural level, and our expectations would have no cause to exist if we did not suppose that they had the faculty to resist natural penchants and thus to avoid allowing a *situation to turn into a determination.*

It is this definition of man—of humans—that Simone de Beauvoir brings to bear in *The Second Sex.* Seeking and examining the conditions under which the emancipation of women could take place, she specifies, with good reason, that such a "problem would be meaningless if we supposed that a physiological, psychological or economic destiny weighed upon women."[7] If we fate women to immanency, if we refuse the possibility of transcending nature, their own nature included, and hold them as tied for all eternity to the domestic life to which biology seems to destine them, we reduce them to an animal state. But as Elisabeth Badinter emphasizes, from a similar perspective, "[S]acrosanct nature is today being manipulated, modified and defied in keeping with our desires."[8] The result is a feminism that is humanist (refusing to confuse humanity and animality), egalitarian (women are no more bound than men to the determinations of nature), and republican (it is by breaking away from the sphere of the particular determinations of nature in general that one rises to the universality of culture and ethics).

This, of course, is a perspective that the philosophies of difference cannot tolerate. It is significant that radical ecology and differencialist feminism have joined forces under the banner of "ecofeminism" to battle this republican existentialism. The polemic, as we shall see, is quite interesting.

But first, what is meant across the Atlantic by "ecofeminism"? The term, coined by Françoise d'Eaubonne, appeared for the first time in 1974.[9] It was quickly adopted in the United States where, as would be expected, it has experienced enormous popularity. What is it? It is the idea—which is simple in principle—according to which there exists a direct link between the oppression of women and that of nature, so that the defense of one cannot be separated from that of the other without harm. Hence the definition offered by one of the leaders of the movement, Karen J. Warren:

> As I use the term, *ecofeminism* is a position based on the following claims: (i) there are important connections between the oppression of women and the oppression of nature; (ii) understanding the nature of these connections is necessary to any adequate understanding of the oppression of women and the oppression of nature; (iii) feminist theory

[7] Simone de Beauvoir, *Le deuxième sexe* (Paris: Gallimard, 1949). [The Second Sex, translated by H. M. Parshley (New York: Knopf, 1971).]

[8] Elisabeth Badinter, *L'un est l'autre* (Livre de Poche), p. 315.

[9] See Sessions, *The Deep Ecology Movement*, p. 115.

and practice must include an ecological perspective; and
(iv) solutions to ecological problems must include a feminist perspective.[10]

At first glance, one might think that such a program would
place ecofeminism close to deep ecology. Are not both movements
based on a mutual rejection of Western civilization after all, with its
single-minded pursuit of domination and mastery? Are they not
both, in this respect, opposed to a reformist-type environmentalism?
These things are true. And yet ecofeminism turns out to be fairly
hostile to deep ecology on one point. According to ecofeminists, deep
ecologists make the mistake of fighting "anthropocentrism in general." What is in question is not the Western world's "human centeredness" but its "male centeredness." A truly "deep" ecology should
surpass the variety that bears this name today—a name too presti-
gious for it—and take into consideration the only truly relevant
question: How can the ties uniting the domination of women and
that of nature *by males* be described? The deconstruction of the humanist tradition, which is of course called for in both cases, cannot be
fully accomplished if one fails to perceive that the critique of anthropocentrism must be replaced by that of *androcentrism*. This is the *leit-
motiv* of this new fundamentalism which we find explained, notably,
in an article by Ariel Kay Salleh bearing the highly significant title
*Deeper than Deep Ecology: The Ecofeminist Connection*:

> There is a concerted effort [in deep ecology] to rethink
> Western metaphysics, epistemology, and ethics, but this
> "rethink" remains an idealism closed in on itself because
> it fails to face up to the uncomfortable psychosexual
> origins of our culture and its crisis . . . But the deep ecol-
> ogy movement will not truly happen until men are brave
> enough to rediscover and to love the woman inside
> themselves.[11]

[10] Karen J. Warren, "Feminism and Ecology," in *Environmental Ethics* 9
(1987).

[11] Ariel Kay Salleh, "Deeper than Deep Ecology: The Ecofeminist
Connection," *Environmental Ethics* 6 (1984), p. 339 esp.

Salleh's thesis, which is representative of that of the entire movement, is that the hatred of women, which ipso facto brings about that of nature, is one of the principal mechanisms governing the actions of men (of "males") and, thus, the whole of Western/patriarchal culture.

It would be wrong, seen from Europe, to think that this is simply a fantasy, one of those hyperboles characteristic of fringe groups, which we ourselves had abundant experience with in the 1960s. For ecofeminism is beginning to occupy a less than negligible place in the heart of American feminism and beyond: it is omnipresent in universities, where it strongly contributes to the reign of intellectual terror exercised in the name of political correctness and the right to be different—the demand for which evolves easily into a demand for a difference in rights.[12] It has already suscitated an abundant literature in which highly diverse positions have emerged based on the common program just outlined.

To attempt to clarify these debates and outline with some precision the ways in which ecofeminism also culminates in a praise of difference, I will first single out three of its fundamental tendencies.[13] They rely on the fact that the origin of and link between the exploitation of women and that of nature can be explained by three at times divergent philosophical positions. The first traces this double oppression to the appearance of *dualism,* the second to that of *mechanistic science,* while the third bases it directly on *difference,* on sexually differentiated personality formation or consciousness.

The explanation by dualism has the merit of being simple. Here is how it is presented in a rather evocative formulation by Val Plumwood, in an article which attempts to synthesize the various aspects of ecofeminism:

> In the ecofeminist perspective, Western thought has been
> characterised by a set of interrelated and mutually reinforcing

[12] Affirmative action is an abstract phenomenon but exerts very real pressure at American universities.

[13] Here I am using the typology proposed by Val Plumwood in "Ecofeminism: An Overview and Discussion of Positions and Arguments," *Australasian Journal of Philosophy* 64, suppl., (June 1986).

dualisms that involve key concepts for understanding the
social structure. Some of these can be set out as follows (the
list is by no means complete):

| SPHERE 1 | contrasted with | SPHERE 2 |
|---|---|---|
| Intellect, mind, rationality, spirit = mentality | | Physicality (body/nature/matter) |
| Human | | Nonhuman, animal, nature |
| Masculine | | Feminine |
| What is produced culturally/historically | | What is produced naturally |
| Production | | Reproduction |
| Public | | Private |
| Transcendence | | Immanence |
| Reason | | Emotion |

Of course it must be added that three further assumptions are
required for these dichotomies to form the root of the exploitation of
the female/nature. First that there exists a hierarchy of the higher
and the lower: the spirit is more worthy than the body, the human
than the nonhuman, the masculine than the feminine, and so on.
Then, that the second sphere should play an *instrumental role* at the
service of the first: man (the male) is thus authorized to use women,
nature, animals, and so on, to his own ends. Finally that a polarity
exists between each side, with the former sphere alone defining what
is considered to be authentically human.

According to ecofeminists, such an interpretive grid has dominated Western thought at least since Plato, and has maintained itself
to our day through the Christian tradition, Cartesianism, the philosophy of the Enlightenment, the liberalism of the Declaration of
Rights, the French revolution, and so on, each of which can be analyzed as a stage in the creation of the prevailing patriarchal ideology.

The explanation of the phenomenon of domination by the emergence, starting in the sixteenth century, of mechanistic science, which broke with the old vision of the world as an *organized cosmos,* seems to be opposed to that invoked by dualism. According to Carolyn Merchant, whose book *The Death of Nature* may be considered a model of the genre, only with the advent of modernity did the identification of the woman with nature become pejorative. For among the Ancients and continuing into the middle ages, according to Merchant, "The root metaphor binding the self, society, and the cosmos was that of an organism . . ."[14] Now, an organism is characterized by *the interdependence among the parts within it,* so that no part can be disvalued without the ensemble as a whole suffering. It is this positive relationship of reciprocity that has been destroyed in modern times.[15]

This breakdown into periods may of course be disputed: from the point of view of the first group of ecofeminists, it leaves out the fact that the dualism of the body and soul was already present among the Ancients themselves, in Plato for instance. But one may also respond that Plato is not representative of Antiquity which, for the most part, it is true, adopted an organistic vision of the universe. Whatever conclusion one may reach in this debate, ecofeminists all agree that the modern world's introduction of a Cartesian-style instrumental rationality only aggravated the dualisms already present in prior traditions. In this sense, the mechanistic thought explanation is not as opposed to the first explanation as is sometimes suggested.[16]

[14] Carolyn Merchant, *The Death of Nature: Women, Ecology and the Scientific Revolution: A Feminist Reappraisal of the Scientific Revolution* (New York: Harper and Row, 1980), p. 1.

[15] "The rise of mechanism laid the foundation for a new synthesis of the cosmos, society, and the human being, construed as an ordered system of mechanical parts subject to governance by law and to predictability through deductive reasoning. A new concept of the self as a rational master of the passions housed in a machine-like body began to replace the concept of the self as an integral part of the close-knit harmony of organic parts united to the cosmos and society. Mechanism rendered nature effectively dead, inert and manipulable from without" (ibid., p. 214).

[16] See on this point Val Plumwood who considers the two theses to be incompatible.

The third position brings us to the heart of our subject. It lies in the attempt to find the origin of the dichotomies in question (spiritual/natural, human/animal, soul/body, and so on) in the *differences* between *genres,* if not between *sexes:* indeed, one mustn't confuse a biological situation with the psychosociologicalcharacteristics that may be ascribed to it by various modalities. As Val Plumwood specifies, it is not biology, but the different experiences produced by different bodies and socialization that determine this aspect of the formulation of genres. In fact, we will have the occasion to see that this nuance, legitimate and desirable in and of itself, is difficult to maintain in practice. What is more, it already tends to blur within Plumwood's own formulation: in saying that different experiences are *produced* by different bodies, is she not suggesting that there exists a biological, and not only a cultural and a historical, *determination?*

As the case may be, the main thesis of this third form of ecofeminism is that the genesis of the double exploitation, women/nature, which we are trying to explain is directly linked to these differences of *genre.* Rather than launching into a general account of this perspective, I will point out several significant examples of the interpretations to which this new version of feminism gives rise.

Let us begin with the essence of the matter, which is to say the concern with rationalism that characterizes masculinity, so it seems, and prompts men to hate the irrationality of natural emotions and sentiments. Where does it come from? According to Rosemary Radford Ruether, whose book *New Woman, New Earth* is authoritative on the subject, this "excessive priority of the rational" comes from men's inability to create life. This insurmountable handicap generates a tragic "effort to deny one's own mortality, to identify essential (male) humanity with a transcendent divine sphere beyond the matrix of coming-to-be-and-passing-away. By the same token, women became identified with the sphere of finitude that one must deny in order to negate one's own origins and inclusion in this realm. The woman, the body, and the world are the lower half of a dualism that must be declared posterior to, created by, subject to, and ultimately alien to the nature of (male) consciousness in whose image man made his God."[17] And it is the inevitable deception, resulting from the fragility of this construction, which "ends logically in the destruction of the earth." QED!

From here on, everything fits together beautifully; analysis by genre applies felicitously to the interpretation of diverse cultural phenomena, including those considered by others, for what must clearly be misguided or dubious reasons, to be highly complex. Should we wish, for example, to fully comprehend the true origin as well as the ultimate significance of Cartesianism, here it is:

> Descartes' extreme reationalism and his subject-object dualism are the products of an extremely masculinist view of self and reality, a view that is shared by many males in modern society. Cut off from their feelings, men become isolated, rigid, overly rational, and committed to abstract principles at the expense of concrete personal relationships. As a result of their attachment to abstract doctrines, males have developed highly rationalistic moral philosophies. Such philosophies include little or no role for *caring* and *feeling* as preconditions for ethics, including the ethic concerning humanity's relation to nature.[18]

Or if we would like to understand the essence of the liberal theories of law and particularly the Declaration of 1789, here too the explanation cuts through the matter like a knife:

> The view of a separate, autonomous, sharply individuated self embedded in liberal political and economic ideology and in the individualist philosophies of mind can be seen as a defensive reification of the process of ego development in males raised by women in a patriarchal society. Patriarchal family structure tends to produce men of whom these political and philosophical views seem factually descriptive and who are, moreover, deeply motivated

[17] Rosemary Radford Reuther, *New Woman, New Earth.* (New York: Seabury Press, 1978), p. 195.

[18] Zimmermann, *Feminism, Deep Ecology and Environmental Ethics,* p. 27.

to accept the truth of those views as the truth about themselves.[19]

Thus, the thrust should not be toward a call for the rights of animals, trees, or rocks, since the very idea of law is disqualified due to its "male" origins. The "legal-ethical extentionism" characteristic of ordinary ecology betrays its "androcentric" roots:

> The doctrine of natural rights is unsuitable for establishing
> a nondomineering relation between humanity and nature
> because it (1) is androcentric, (2) regards nonhuman beings
> as having only instrumental value, (3) is hierarchial, (4) is
> dualistic, (5) is atomistic, (6) adheres to abstract ethical
> principles that overemphasize the importance of the isolated
> individual, (7) denies the importance of feeling for inform-
> ing moral behavior, and (8) fails to see the essential related-
> ness of human life with the biosphere that gave us birth.[20]

It should be added that these partial or "regional" analyses of multiple facets of modern culture are rooted in a more general frame-work, in which surprising derivatives of phychoanalysis, sociology, or political science are jumbled together, willy-nilly. Sometimes the en-tire history of humanity itself is summed up in a few lines, as in the following representative example, which I will also quote to give an idea of the strangeness of this literature:

> This male view of the self as an isolated ego stems from
> early childhood relations between son and mother. At first,
> the little boy identifies himself with his mother; later,
> however, he discovers that he is sexually differentiated from
> her. Seeking to gain his own sexual identity, the boy expe-
> riences his own withdrawal from his mother as abandon-
> ment. Experiencing profound anger and grief because of
> this perceived abandonment, he subsequently fears, mis-
> trusts, and hates women . . . His fear and anger leads him

[19] Naomi Scheman, "Individualism and the Objects of Psychology," in *Discovering Reality* (Boston, 1983).

[20] Zimmermann, *Feminism, Deep Ecology and Environmental Ethics,* p. 34.

to want to dominate both the woman (mother image) within himself and the woman outside of him . . . 'Mother Nature' then appears as a threatening, unpredictable force from which a man must differentiate himself and which he must control.[21]

This, then, is the primary configuration that explains, according to this paranoid vision of history, the emergence of all Western civilization, with its plan for world domination and its renowned hatred of women. Two conclusions can be reached, which serve to specify the redemptive role of ecofeminism.

The first is that salvation, in effect, can only come from women. Only women are in a position to escape the threat brought upon humanity as a whole by the presence of malefic dualisms, and this is for one good and simple reason: unlike men, they have not broken off from nature. Their experience of their bodies, their attachment to the natural mechanisms of life are far too present, far too strong for them to dream of "emancipation" from them. According to Elizabeth Dodson Gray, the cycle of reproduction provides "an inescapable limit upon her physical existence. It would be difficult for such a woman to dream up a sense of herself as unlimited or as all-conquering mind or as a Promethean self."[22] From the same perspective, Mary O'Brien endeavors to show how the "reproductive conscience" of the woman is the experience of a fundamental continuity with biological rhythms based on the fact that "she herself was born of a woman's labor, that labor confirms genetic coherence and species continuity," unlike the reproductive conscience of males which is splintered and discontinuous. The process of reproduction is thus "a synthesizing and mediating act [which] confirms women's unity with nature."[23] An authentic ecological discourse must, therefore, draw inspiration from this symbiosis. For only a oneness with nature, which is *natural in itself,* can guarantee true respect for it. Hence the distance it is fitting

[21] Here Zimmermann, p. 31, summarizes the arguments of Nancy Chodorow, author of *The Reproduction of Mothering* (Berkeley, 1978).

[22] Quoted by Val Plumwood, p. 125.

[23] Mary O'Brien, *The Politics of Reproduction* (Boston: Routledge and Kegan Paul, 1981), p. 59.

to maintain from all forms of ordinary ecology, including those that appear the most radical.

But we must also note the distance that separates such a world-view from the traditional faces of feminism. Beginning, of course, with existentialist feminism. I mentioned previously how this feminism subscribed to the egalitarian and humanist view of emancipation with respect to natural determinisms. From the point of view of ecofeminism, this supposed "emancipation" can only be a disappointment, the ultimate disappointment, in truth, since it implies a simultaneous negation of femininity and of naturality in favor of a typically masculine model of freedom. This is why Val Plumwood, drawing inspiration from Mary Midgley, reproaches Simone de Beauvoir, in terms which should also be quoted:

> Thus for Simone de Beauvoir woman is to become fully human in the same way as man, by joining him in distancing from and in transcending and controlling nature. She opposes male transcendence and conquering of nature to woman's immanence, being identified with and passively immersed in nature and the body. The "full humanity" to be achieved by woman involves becoming part of the superior sphere of the spirit and dominating and transcending nature and physicality, the sphere of *freedom* and controllability, in contrast to being immersed in nature and in blind uncontrollability. Woman becomes "fully human" by being absorbed in a masculine sphere of freedom and transcendence conceptualised in human-chauvinist terms.[24]

Here we can measure the distance from existentialist feminism: it is by affirming her *difference* from "males," by insisting instead on her *specific* proximity to nature, that the woman, like the proletariat in days past, incarnates the redemptive portion of humanity. The danger inherent in such a position is obvious. Simone de Beauvoir had already foreseen it and Elisabeth Badinter has analyzed it: an insistence on the "naturality" of women threatens to revive the most time-worn clichés about "feminine intuition," the vocation of mother-

[24] Plumwood, p. 135.

hood and the irrationality of what could well, under such conditions, pass for the "second sex." To assert that women are more "natural" than men is to deny their freedom, thus their full and whole place within humanity. That the ecofeminists hate Western civilization and modernity is their business. That they wish to find natural justifications for this hatred means playing the game of biological determinism, of which all women will suffer the consequences if it is to be taken seriously. The demand for the right to be different ceases to be democratic when it becomes a call for a difference in rights.

## seven

## Democratic Ecology and the
## Question of the Rights of Nature

Under its two profiles, deep ecology presents an unsavory face to the democrat. It also poses just as serious challenges to the humanist ethics it claims to surpass.

The first challenge is political. At a time when the mourning of utopias seems to be popular even among those who had never cared for them to begin with, it opens a new space for action and reflection. It has a little of everything: science and morality, epistemology and philosophy, cosmology and mysticism. Enough to open new horizons to a militant corps lacking reasonable places to invest its energies. But there is more: deep ecology poses *real questions* which the critical discourse denouncing the stale smell of fascism or new leftism cannot brush aside. No one can convince the public that the ecology movement, radical as it may be, is more dangerous than the dozens of Chernobyls that threaten to erupt. And say what you will about the foolishness of the antimodern themes debated by the new fundamentalists, maintaining a "laissez-faire" liberal attitude is nonetheless insane. That it will ultimately be by means of advancements in science and technology that we manage one day to resolve the questions raised by environmental ethics is more than likely. Yet to imagine that the solutions will appear on their own, as if part of the natural evolution of things, without our having to mobilize collective thought and action, is childish. Hence the necessity of integrating ecology within a democratic framework—it is because this is too serious a matter for the deep ecologists to handle alone, and not because their questions are automatically invalid, that they must not be left with a monopoly on the discussion.

Particularly since no one can remain indifferent to a questioning of the liberal logic of production and consumption. In one form or another, romantic or utopian, this questioning has never ceased to accompany the rise of democratic societies, and it would be naive to imagine that for want of combatants, communism being dead and fascism disabled, the battle would vanish as if by magic. The end of history, whatever one may think, has not arrived. On the contrary: alone with ourselves at long last, saddled with a "democratic melancholy,"[1] we too are in a position to measure the dissatisfactions of the consumerist dynamic. Without getting too religious, one suspects that man is not on this earth to buy higher and higher performance cars and televisions; though our final destination may remain a mystery, this, certainly, is not the ultimate goal. Fundamentalism, whether political or religious, can always count on this anxiety: it is an inherent part of the fate the neutral liberal State reserves for individuals, left to themselves and deprived of the salvation of a strong collective ideology.

But it is also on a metaphysical level that deep ecology scores points and poses a second challenge to humanism. For the two dominant forms of anthropocentrism, Cartesianism[2] and utilitarianism, fail to do justice to the feeling for nature that prevails today. From the simple point of view of our immediate "intuitions," even of our carefully weighted convictions, we cannot entirely dismiss our impression that nature possesses a certain value *in and of itself,* that it tends at times to surprise us, even to instill wonder in us, outside of any consideration of mastery or utility. This "lived experience" seems immediate, independent, at least, of our goals. As if nature concealed both the worst tendencies—the reign of brut force and violent death—and the best, harmony and beauty being the most visible signs. The status of such a sentiment, of course, has still to be

---

[1] Pascal Bruckner justifiably resurrects a major theme of Nietzsche and of Carl Schmitt when he associates this melancholy with the disappearance of "enemies."

[2] Even though Cartesianism cannot be reduced to the caricature often given of it. See on this point, as well as on the critique of Michel Serres's theses, the book by François Guéry and Alain Roger, *Maîtres et protecteurs de la nature* (Champ Vallon, 1991).

evaluated, as if the theses at war in *the conflict between ecocentrism and anthropocentrism ultimately never manage to define the basic facts of the problem. One accords too much to nature, the other too little, each side finding solace, as in any opposition, in the adversary's weaknesses.* There is no question that one of the criteria defining the project of a nonmetaphysical humanism is the ability to understand the stakes of such a conflict, which is basically analogous to the one we examined concerning the rights of animals.

The two orders evoked, the political and the metaphysical, meet here. For the affirmation of the rights of nature, when it takes the form of the latter's being instated as a legal subject, implies the rejection of a certain type of democracy—a democracy inherited from the Declaration of Rights and which has inscribed itself in our liberal-social-democratic societies. Which is why deep ecology is at least coherent when it claims to be "deconstructing" modern humanism and the liberal world, one of its political expressions. The idea that one could "add" a "natural contract" to the Declaration of Rights is not very valid philosophically. Clearly, there is some discontinuity between the two contracts: within the framework of legal humanism, nature can occupy only the status of *object,* not of *subject.* This in fact is why the Green parties refer to themselves as "revolutionaries" and call for a true conversion. To use a metaphor dear to their hearts: he who wishes to go to Marseille and mistakenly finds himself on the road to Lille will not simply slow his pace. He'll have to turn around! Basically: we don't need reforms, we need a revolution!

If we do not wish to simply disqualify deep ecology by harping on its excesses and dangers, if we also wish to distinguish what might turn out to be relevant, we must consider these questions: Is it necessary, to ensure the protection of our environment, that we grant it equal, even superior, rights to those granted human beings? To what extent and how can we speak of the "rights of nature"? Does the fact that we recognize nature's dignity imply the radical deconstruction of humanism *in all its forms?* Would an internal critique of this anthropocentrist tradition enable us to do justice to the concerns of radical ecology without renouncing democratic principles? And conversely: How can political liberalism, the liberalism of the rights of man, integrate the preoccupations of environmental ethics?

In the end, we must dare to put forth propositions that aim to define the theoretical and practical frames of reference necessary to defend a democratic ecology. Such a program could initially be defined as follows: neither Cartesian nor utilitarian anthropocentrism nor deep ecology. How, working within this "neither nor" framework, can we meet the challenge within the two areas mentioned: the political and the metaphysical?

## An Internal Critique of Fundamentalism

In constructing the ideal-type of deep ecology, I had occasion to point out the perverse effects of this new fundamentalism that were most apparent: radical antimodernism leads to a fascination with authoritarian political models; moral scientism ineluctably leads to dogmatism; the divinization of nature implies a rejection of modern culture, suspected of causing man's uprootedness; praise of diversity positions itself as hostile to the public republican space; and so on. But the perception of these troubling tendencies remains *external to its object*. The critiques they suggest will seem pertinent only to those who already situate themselves within a democratic perspective founded on an acceptance, however partial, of the principles of political (if not economic) liberalism. For the others, whose goal is precisely to bring an end to this democratic-liberal universe, they will seem unconvincing, even automatically invalid, since they issue from the framework of thought from which the goal is to break away.

It is, therefore, necessary to go one step further, to discuss the *internal* critiques of deep ecology, that is to say the objections rooted in the *inherent* difficulties of such a position.

The first leads us back to a paradox which, though simple in appearance, is nonetheless difficult to surmount: while the deep-ecology program rests entirely on the rejection of anthropocentrism (Cartesian or utilitarian) in the name of the rights of the ecosphere, the logic of their own reasoning causes them to fall back on one of the most absurd forms of *anthropomorphism*. Philip Elder, one of the supporters of a "superficial" and "environmentalist" ecology, has formulated this major difficulty in amusing, but fundamentally incontestable, terms: discussing Stone's theses on the rights of trees "populating" the valley of

Mineral King, he mentions that his staunchest adversaries "always suppose that the interests of objects (mountains, lakes and other natural things) are *opposed to development*. But how do we know? After all, isn't it possible that Mineral King would be inclined to welcome a ski slope after having remained idle for millions of years? . . . Aren't the deep ecologists acting 'anthropocentrically' themselves when they claim to know what is best for the natural environment?"[3]

Indeed: to say that animals have "interests" is already debatable, even if we recognize in them the capacity to feel pleasure and pain; But doesn't claiming the same for trees, rocks, or the biosphere as a whole lean toward an animism comparable to that which informed the medieval trials of grasshoppers or weevils? There is an insurmountable logical error at the very foundation of fundamentalist reasoning. This error has a name: "performative contradiction," the model for which is furnished by the following type of proposition: "I was on a boat that sank and there were no survivors." The *content* of the statement contradicts the conditions of its enunciation. This discordance can be found in the legal arguments of deep ecologists: imagining that good is inscribed within the very being of things, they forget that *all valorization, including that of nature, is the deed of man and that, consequently, all normative ethic is in some sense humanist and anthropocentrist.* Man can decide to grant a certain respect to nonhuman entities, to animals, national parks, monuments, or cultural works: whether we like it or not, the latter always remain *objects and not subjects of law.* In other words, the idea of creating a normative, antihumanist ethics is a contradiction in terms. In wishing to maintain the idea of value yet suppressing the conditions under which it becomes possible, the fundamentalists fall into the performative contradiction: they forget that it is *they, as human beings, who value nature, and not the reverse, that it is impossible to disregard this subjective or humanist moment and project into the universe itself an "intrinsic value."* No doubt there are aspects *of nature* that *move us*—a phenomenon which deserves to be described and analyzed against a certain Cartesianism—but that doesn't mean that it is possible to disregard the

[3] Philip Elder, "Legal Rights for Nature: The Wrong Answer to the Right(s) Question," *Environmental Ethics* 2:111.

"us." On the contrary, it is precisely by attempting to disregard subjectivity that the philosophy of nature leads to the illusions of
anthropomorphism. In opposition to Jonas's thought, we must remember that if the end wishes to be "moral," it can never "reside in
nature," that while the "biological foundation is necessary," as Ricoeur reminds us, it "ceases to be sufficient" when it comes to determining the conditions, not of the simple survival of men on earth,
but of a *good* life—which is an entirely different matter.[4]

One may object that the idea of a law of nature is just a literary
metaphor, destined to arrest the attention of a public sunk in lethargy.
Does this mean Michel Serres does not seriously consider that man
and nature can enter into a genuine "contract" together, in which they
would consider themselves, as he suggests, equals? Perhaps. But what
is the point of such poetic license if one must immediately discount
its philosophical weight? What relationship remains between the social contract and its counterpart in nature if the latter is not a true
pact but a short article appended to the first, a hastily added frill?
Why use such a strong image if its meaning must then be negated, if
it does not represent a genuine concern for transforming the beings of
nature into *legal subjects?* And what difference then remains between
this supposedly new vision of our relationship to nature and that of
the "superficial" ecologists and "anthropocentrists?"

Instating nature as a legal subject, deep ecology, when strict,
turns the material universe, the biosphere, or the *Cosmos* into an ethical model to be imitated by men. As if the order of the world were
good in and of itself, all corruption emanating from the polluting
and selfish human species. I have already suggested that such a romanticism led to the repudiation of the best of modern cultural,
whether of law, which won out over the natural rule of force, or of
the heritage of the Enlightenment and the French Revolution,
which won out over the reign of tradition and "natural" facts. But
there is more, if we look at things from the perspective of an internal critique, which we need to do faced with those propelled by a
hatred of modernity: it is that the sacralization of nature is *intrinsi-*

[4] See Ricoeur, *Le Messager européen,* no. 5, p. 217.

*cally* untenable. Like religious fanatics, hostile to all medical intervention because they deem it contrary to divine intentions, deep
ecologists blithely disregard all that is hateful in nature. They retain
only the harmony, the beauty and the peace. It is in this spirit that
some eagerly disqualify the category of "pests," considering that
such a notion makes no sense, being entirely anthropocentrist.
Drawing inspiration from theology, they suppose that nature is not
only the supreme Being but also the *ens perfectum,* the perfect entity
that it would be sacrilegious to attempt to modify or improve. But
what about viruses, epidemics, earthquakes, and all that we justifiably refer to as "natural catastrophes"? Can we say they are "useful"? To whom and to what? Should we consider that they possess
the same legitimacy as do humans to persevere in their existence?
Then why not propose a right of cyclones to devastate, of earthquakes to engulf, of microbes to infect the sick? Unless we adopt an
anti-interventionist attitude in every way and in all cases, we are
forced to admit that nature taken as a whole is not "good in and of
itself" but that it contains both good and bad. From whose point of
view? one may ask. From man's, of course, who remains until proof
to the contrary the only being able to make value judgments and, as
the Bible says, to separate the wheat from the chaff. It is a matter not
of denying that nature, *in and of itself,* can be beautiful, useful, or
even "generous" (which again brings up the question of the limits of
Cartesianism), but only of emphasizing that it is not voluntarily and
consistently so, as would be the divinity in which we are asked to
believe, and that it is always we, human beings, who must ultimately judge. As in economics, the philosophy of nonintervention
presupposes the sacralization of a natural world harmony. This is a
metaphysical, even a mystical optimism for which we unfortunately
have no justification.

   Man can and must *modify* nature, just as he can and must *protect*
it. The philosophical question of rights inherent to natural beings rejoins the political one of our relationship to the liberal world. Within
these two orders, in which a nonmetaphysical humanism and an anti-
Cartesian anthropocentrism enter into competition with deep ecology, it is necessary to clearly explain one's choices. No doubt they

would include praise of an internal critique and the acceptance of indirect duties toward nature.

## Democratic Policy and the Choice of the Internal Critique

The external critiques of the liberal universe, those made in the name of a radically different other world, whether past or future, risk leading again to the worrisome seductions of the various totalitarianisms. The internal critique, the critique of democracy, real and imperfect as it is, in the name of its promises and its own principles, is, by definition, the only critique that remains compatible with the requirements of democracy. It nonetheless comes up against a difficulty upon which religious or political fundamentalisms rely: after two centuries of messianic utopias, the conversion to reformism seems rather unexalted, too tame, too flat to seduce militants, which the death of communism and leftism have left in a state of shock. A difficulty that is all the more delicate to surmount as the feeling of mourning is far from incompatible with the lucid acceptance of past "mistakes" and the resolute abandonment of defunct theories. It is one thing to recognize the objective misdeeds of communism, it is another to definitively bury militantism, the associative life and, more profoundly still, the perspectives it opened onto the classical question of the meaning of existence. That Marxism was a "religion of earthly salvation" is no longer in question. Yet we need to fully understand the exact signification of this statement if we wish to understand why new fundamentalisms, beginning with those that propel deep ecology, are hovering around the spoils to take succession and quickly encumbering the chances of a democratic political structure.

The feeling of emptiness that comes over former faithfuls seems to me, essentially, to be based on a philosophical and historical misunderstanding of the relationship between politics and religion, a misunderstanding which continues to prevent the reformulation in positive terms of the principles of *"radical reformism."* What we have been experiencing, in effect, for more than two centuries now, is the history of a slow but ineluctable dissociation of these two spheres, formerly so intimately linked that believers could move imperceptibly from one to the other. To put it simply, in the wake of the French Revolution, we have experienced a double break with religion.

The first is so essential it can be considered responsible for the creation of the European cultural sphere as a whole. We are talking now about the birth of secularity; as Marcel Gauchet emphasizes in *Le désenchantement du monde,* what is specific to our democratic spaces, more than any other feature, is that norms and collective values are no longer rooted in a theological universe. This is what continues to separate us from "Islamic" republics. It is this event that the jargon of contemporary philosophy designates as "the end of theologico-politics." Beyond its words, which are burdened with the weight of tradition, the "Declaration of Rights" symbolizes the advent of norms which, though still collective in their vocation, no longer draw legitimacy from religious inspiration—in principle, at least, they must derive solely from the will of individuals, whether they are authors of the law directly, as Rousseau and his disciples would have preferred, or whether they express themselves, as in the representative system, by intermediary of their parliament. Another way of saying it is this: men have discovered that they can and even must resolve the questions of what constitutes a good life or good decisions *on their own,* without taking orders from above. We no longer possess a shared body of wisdom, the collective markers that in the past could be found "ready made," so to speak, in a shared religion (though in truth they were entirely *manufactured* by the clerical authority and the history of interpretations . . .).

The second break is more recent. It coincides with the crumbling of communism in all its variations. While the necessary distance may still be lacking for historical analysis, we can nevertheless hypothesize as to the effect of a disappearance so considerable it leaves nothing intact on the order of politics. Like it or not, liberals or social democrats, deprived both of an enemy *and* of an ally, will not emerge from the upheaval unscathed. One point of clarification is required as to the meanings of the word religion. It can be understood in three ways: first as a common *tradition,* imposed on men *from without* and originating with a divinity, to bind them among themselves and ensure their social connection on the theologico-political model; second, as a discourse of *superstition,* which is to say of man's alienation—it is from this perspective that the materialists of the Enlightenment, and later their Marxist or Freudian heirs, denounced faith as the "opiate of the people" or the "obsessive neurosis of humanity"; third and finally, as the place for questioning the meaning of existence, or, if one prefers, "of the meaning of meaning" (of the ultimate and supraworldly significance of all partial and intraworldly significance).

The first rupture between politics and religion was that of secularity, which broke with *tradition:* it stands as the supreme act of the Revolution against the order inherited from the Ancien Régime. It is within this secularity that the belief in a God tended to become defined as superstition. The second break, on the other hand, was directed against the third meaning of the term religion. It is at the origin of this feeling of emptiness which, though understandable, nonetheless seems illusory to me. Emancipated from the tutelage of religious authorities, freed from dogmatic partisan lines, individuals seek the meaning of their existence *outside of religion and politics.* Meaning is now situated in the present, or at most in the immediate future, but the past and the future, *which come before or after life,* no longer figure much in the picture. We "exist" almost like a *project,* constantly setting all sorts of "goals"—professional, romantic, cultural, or other. Within these *small schemes,* which are like so many self-contained bubbles, our actions take on meaning. But the question of the meaning of these projects, or, if one prefers, the question of the meaning of meaning, can no longer be posed *collectively within the heart of a secular universe.* And this diminishment is valid both for religion, which is becoming a private affair, and for the utopian politics of days past. To think that constructing a new Europe or providing economic aid to the countries of the East will be an adequate substitute, in the hearts of militants, for the dream of building a free and classless society is a blunder which, omnipresent though it is among politicians, is nonetheless colossal. Moreover, the intellectuals and journalists who, page after page, deplore the disappearance of hard and fast ideological cleavages, the good times "when of course we made mistakes, but we did it with passion," are victims of the same illusion, having failed to see that the depression is in no way temporary. We are not *passing through,* as Edgar Morin believes, a "period of mythological low waters," a temporary retreat into the private sphere and egotistical self-interest, soon destined to be replaced by the emergence of a new master plan (ecology, of course!). Rather, to all appearances, the crisis, here as elsewhere, is structural, "historial" if one prefers, which is to say tied in with the adult development of the secular and democratic universe.

This diagnosis is not just the result of an abstract and disembodied hypothesis. It is based also on several observations. Indeed, there is every indication that the question of the meaning of existence has receded from religious politics and moved into other spheres: those of ethics and culture, *understood as the blossoming of the individual.* The proof being that, in the course of the 1980s, the only "new" political movements were not political movements at all! Moral and cultural in nature, they have emerged in "civil society," or at most on the margins of traditional parties. This is true of antiracist organizations, which claim to rely on a "youth culture" (which is essentially musical) and on moral imperatives, beginning with the battle against exclusion, the demand for respect of the "other." It is true as well, paradoxical as it may seem, of the National Front, which also leaves behind classical politics to defend "values"—the "moral rearmament of France" and a cultural scheme centered around the defense of national identity. But it is true too of the impressively successful humanitarian organizations, and, last but not least, of ecology—which seen from this perspective does not strike me as directly descended from a renaissance of leftist and worker movements, as Alain Touraine seems to believe.[5]

This, I believe, is the reason for the error of those who still mourn the fall of the revolutionary ideal and associate reformism with a lukewarm, colorless, and tasteless brew: having failed to understand that, in a secular democracy, politics had to leave the bosom of religion, they see only the bad side of history—especially since it most often coincides with their own youth, a simple question of fact

[5] See Touraine's comments in dossier no. 11 of the *Nouvel Observateur* (1992).

but one that is important nonetheless. In truth, if the democratic space is indeed one that incorporates an internal critique, they should, since they call themselves democrats, celebrate this liberating and salutary mourning, as, *for the first time no doubt in the history of humanity, we are living in a time when this critique, which the eighteenth-century Aufklärer were already calling for, has reached the minimal threshold of maturity.* Reformism is not the model we must be content with for want of better when revolutionary hope fails, but rather the only position consistent with leaving the world of childhood. Not only is it the only attitude compatible with the democratic rejection of partisan lines and dogmatic authority, not only does it cease to hold out the mystical hope of fighting for something beyond the real world, but unlike revolutionary ideology, which is oriented toward a final goal, it opens an *infinite* space for reflection and action.

And it is through this idea of infinity—of a task irreducible to the success of an ultimate goal such as the taking of power through revolution—that the secular universe attempts to reappropriate the question of the meaning of meaning. In all the areas in which the heritage of the Enlightenment has had a decisive impact, the notion of infinity has taken hold: whether in science, education, culture, or even ethics, the theological idea of a final objective has lost its sway. We know that we are dealing with infinite processes, that *progress*—ridiculed by radical ecolgists—cannot consist in reaching a final stage, beyond which there are no more questions. *And it is in light of this very notion of infinity that the human being, now defined by his perfectibility, reinterprets the question of meaning.* Indeed, why would politics escape the logic of secularization, which has reached all other spheres of culture?

Are the days of prophets, when the use of intelligence was limited, at times, to the choice of a "camp," to be regretted? The most simplistic divisions—for or against revolution, capitalism, alienation, "symbolic force," self-management, and so on—were enough to separate the good from the bad without any further examination of the issue being necessary. One could speak of the democratization of teaching, of nationalization, of zero growth or the respective roles of the US or the USSR, in complete ignorance of the most elementary facts. The struggle of ideas being the pursuit of the class struggle by other means, everything was considered a pretext for expressing the ethical choices of strategic logic. A sinister time, in truth, when the divisions between intellectuals, true professional ideologues, and experts riveted to their administrative careers enabled everyone to avoid the decisive questions.

Some say that the end of these theologico-political quarrels mark the advent of an era of management in which politics are just another technical skill. This may be overly pessimistic. For the ideal of the internal critique calls for a "master plan," not a substitute for those of years past but perhaps the first one suitable for adults. If we wish to revalorize politics as an autonomous sphere of collective decision, a counterpoint to the ethics and culture of the self to which many today are retreating, we must reformulate the principles of democratic reformism, situating it in the line of the disappearance of theologico-politics. Which unquestionably presupposes a redefinition of the role of politicians and intellectuals: rather than being called upon to again furnish "great messianic masterplans," they will have to help to organize, clarify, and untangle the great debates, the absence of which is becoming untenable to young citizens coming into adulthood. This is the kind of innovation we need.[6]

## Duties toward Nature

The two major problems encountered by deep ecology in its plan to instate nature as a legal subject, capable of being party to a "natural contract," can be further summarized as follows: the first, which is shockingly obvious, is that nature is not an *agent,* a being able to act with the *reciprocity* one would expect of a legal *alter ego. Law is always for men,* and it is for men that trees or whales can become *objects* of a form of respect tied to legislation—not the reverse. The second problem is less obvious: if we accept that it is possible to speak metaphorically of "nature" as a "contractual party," it will still be necessary to specify what it is, *in nature,* that is supposed to possess intrinsic

[6] On the birth of institutions whose mission is to organize democratic debates, see the autumn 1990 issue of the journal *Pouvoir,* devoted to bioethics.

value. The most common response among fundamentalists is that it is the "biosphere" *as a whole, because it gives life* to all beings, or at the very least allows them to sustain their existence. But the biosphere gives life both to the AIDS virus and to the baby seal, to the plague and to cholera, to the forest and to the river. Can one seriously claim that HIV is a subject of law, equal to man?

This objection does not aim to legitimize Cartesian anthropocentrism *a contrario,* only to bring to light how difficult it is to speak of the objective world in terms of subjective rights: How can we move beyond the antinomy of Cartesianism (which tends to deny creatures of nature any intrinsic value) and deep ecology (which considers the biosphere to be the only authentic subject of law)? There is no doubt that this question, in one form or another, will occupy center stage in ecological debates in the years to come. It will be at the heart of philosophical preoccupations with the new status of relations between man and nature, as well as of the legislative projects that will inevitably come to light in industrialized countries. Without aiming to resolve this question, it is already possible to indicate a path for reflection, analogous to that evoked with respect to animal rights.

If animals were only machines, as Cartesians would have it, the question of their rights would *never* have been raised. What may arouse feelings of *obligation toward them,* beyond compassion and pity resulting from plain old *sympathy,* is the non*mechanicalness* of the life they incarnate. Not that we wish to totally disqualify the *sentimental* approach to the question of rights, but rather to seek, beyond simple phenomenological descriptions, the eventual principles of legitimacy. For sympathy is only a *fact* that collides with other facts and, as such, is not a justification. There are those who love bullfighting de facto, those who condemn it de facto, but if we wish to decide the matter de jure, we must rise above the sphere of facts and seek *arguments.* One argument I have already mentioned and which runs counter to those both of the Cartesians and of the utilitarians, justifying the idea of a certain respect for animals, is that they seem to belong to an order of reality that is neither that of stones nor of plants, while not belonging to humanity proper either. Although prompted by the code of instinct, and not by freedom, they are the only beings *in nature* which seems capable of acting according to the representation of goals, thus in a conscious and intentional fashion. It is on this basis that they distance themselves from the realm of the mechanical and grow nearer, by analogy, to that of freedom. They are not simple automata, and their suffering, to which we cannot and even must not remain indifferent, is one of the visible signs of this. Others we can mention include the devotion, affection, or intelligence they sometimes exhibit. In short, it is *as if* nature, in animals, tended in certain circumstances to turn human, *as if it automatically fell into line with ideas we value when they are manifested in humans.*

The meaning of this "as if" must be clarified: it is the indication that the value judgment with respect to animals and to their eventual rights is neither entirely "naturalistic" (as in deep ecology) or entirely "anthropocentrist" (as in Cartesianism and, in certain respects still, Kantianism). For it is *nature itself* that signals toward ideas that are dear to us and not we who project them onto it: contrary to what Cartesians think, it seems reasonable to admit that the cries of animals who suffer do not have the same significance as the ticking of a clock, that the fidelity of a dog is different from that of a watch. Hence the feeling that nature does possess the *intrinsic value* upon which *deep ecologists* rely to legitimize their antihumanism. But from another angle, and this is what they are missing, it is *the ideas evoked* by nature that bestow value upon it. Without them, we would not accord the slightest value to the objective world. What is more, because nature flies in the face of such ideas, because it also produces violence and death, we instantly rescind the high appraisal we made of it a moment earlier when it seemed beautiful and harmonious, or even, in the form of an animal, intelligent and affectionate.

Which shows us why the question of the "rights of trees" should be reformulated separate from Cartesian anthropocentrism (since it is *nature* that evokes the ideas we love) and from fundamentalism (since it is still *the ideas,* and not the object as such, that are the basis for value judgments *which only men are capable of formulating: ethical, political or legal* ends never "reside in nature," which knows no moral finality). We, therefore, must do justice to the feeling that nature is not devoid of value, that we have duties toward it, though it is not, for all that, a legal subject. It is also from this perspective that we could attempt to define *what it is, in nature itself,* that must be respected and what must instead be combated through a well-planned interventionism. Without such a distinction, the idea of duties toward Nature loses its meaning, since it is clear that not everything in nature deserves to be protected equally.

Aside from freedom, traces of which we perceive in the suffering of living beings insofar as it is evidence of a nonmechanical nature, there are two ideas we value and which, therefore, also valorize nature when nature, by chance, happens to "present" or "illustrate" them: those of beauty and finality.[7] Some may prefer artistic and spiritual beauties to those of nature, in keeping with Hegel or the French classics. The romantic attraction to wild virginity is nonetheless so widespread a phenomenon that it must have its reasons. We know that Leibniz, who was a great scholar and wit, liked to return the beetles he examined under his microscope to freedom. The geometric designs of unequaled purity, the infinite richness and harmony of the colors decorating their wings, seemed to him a sign of favor, the generous gift of a material universe turned artist and mathematician for the occasion. For nature is beautiful when it imitates art; this judgment is not anthropomorphic but merely *recognizes* the mystery of natural beauty, that strange phenomenon by which the world, though objective and foreign to us, makes itself in some ways more human than we would have hoped. The harmony of nature is the moment when chaos becomes order *without human intervention.*

An analogous sentiment, I believe, is triggered by the observation of natural finalities. Ecosystems are better designed than most human constructions. As a result, any intervention on our part most often turns out to be extremely damaging; as in the realm of the economy, great caution is required. Even when his intentions are good, man is constantly bringing about unexpected results, "side effects." In France, we "destroy" the foxes because they have rabies. But the population of rodents upon which they prey grows so large that further intervention is required to reestablish an equilibrium, no doubt bringing about additional complications. Such examples are legion. We cannot use them to prohibit taking any action on the world. But

[7] According to a structure which philosophy designates as a "reflexive judgment."

they should at least remind us of *phronesis,* that celebrated caution of the Ancients which is so lacking in our modern politics. Most of all, they indicate what it is in nature that must be respected: in the finality it demonstrates, it often proves superior to us *in intelligence.* This statement is not anthropomorphic either, even if the notion is willfully provocative. Rather, it is concerned with recognizing and, if possible, with preserving what already *appears* human in nature and thus connects with the ideas that are dearest to us: liberty, beauty, and finality.

It is along these lines, distinct from Cartesianism, utilitarianism, and fundamentalist ecology, that we must develop a theory of duties toward nature. Not that nature should be the subject of and party to a natural contract—which seems fairly meaningless. But the *ambiguity* of the enigmatic nature of certain beings cannot leave those of us who care about the ideas they incarnate indifferent. The word ambiguity is apt here: these *mixed* beings, *syntheses of raw material and cultivated ideas,* participate equally in nature and in humanity. Thus a phenomenology of human signs *in* nature must be created to obtain a clear awareness of that which can and must be valued in it. By imposing limits on this basis to the interventions of technoscience, democratic ecology will meet the challenge posed in both the political and metaphysical realms by its fundamentalist competitor.

## The Ecological Sensibility and Democratic Passions

A philosophical program of this sort is far from disconnected from what is sometimes called an "ecological sensibility." For *deep ecology,* despite its seductive sides, despite the media's attraction to radical organizations such as *Greenpeace,* is marginal relative to the heightened concern for the environment prevalent today in all industrialized countries. The love of nature strikes me as being essentially composed of *democratic passions shared by the immense majority of individuals who wish to avoid a degradation in their quality of life;* but these passions are constantly being *claimed* by the two extremist tendencies—neoconservative or neoprogressive—of deep ecology. This is one of the functions of the Green parties, in their fight against reformism. It would, therefore, be wrong to denounce ecology *in*

*general* as "fascist" or "leftist": this would be to miss a major phenomenon, the meaning of the groundswell that is occurring today in
democratic societies, and is not necessarily linked to the renaissance
of romantic nostalgia or utopian messianism.

What is more, it is fairly simple to distinguish the themes that
point to such an interpretation and provide evidence of the presence
of democratic passions in the ethics of the environment. To correctly
understand them, we must remember that ecology formed into political movements toward the end of the 1960s, simultaneous with the
appearance of the student revolts that would culminate in May 1968.
The *Stimmung,* the intellectual and moral atmosphere of this period,
is marked by the emergence of an "ethics of authenticity."

This ethics consists, on the one hand, of rejecting *aristocratic* values, of combating hierarchies in the name of the *egalitarian* and, in
this sense, democratic idea that *all practices are equally valid,* that each
of us possesses the right to live out his difference, to be himself. It is
from this perspective, for example, that the "sexual liberation" movement would reject the traditional discrimination between "normal"
and "deviant," attempting to destigmatize homosexuality and, more
generally, all behavior previously condemned in the name of a normative ideal that arranged all forms of life within a *hierarchy.*

On the other hand, the ethics of authenticity endeavors to disqualify the moral notions of *duty* and *merit:* if it is now "forbidden
to forbid," it is because the transcendent norms, the "ascetic ideal"
Nietzsche denounced in Christianity and in Protestant rigor, no
longer carry weight. If we have any duty, it is to "be ourselves"; if
three is a new norm, it says that *each of us must invent his own norm.*
Previously, ethics consisted in *endeavoring* to achieve, most often
against the grain of one's egoist penchants, the attainment of *standards* external to ourselves. It supposed an effort of the will impelled
by *imperatives* expressed in the form of a "must." Now its goal is the
realization of the self in the idea that the law, rather than being imposed upon us from without, is immanent within each individual.

This democratic and authenticity-oriented individualism is found
in the desire to preserve the environment. For it is against this ideological backdrop that the "concern for the self" emerged, the will to
be "healthy in mind and body." With all due respect to the former

combatants, May '68 was more of a "health" movement than a revolutionary one—what's left of it today is more the "be yourself" ideology, the desire to nourish and develop one's body and mind, than the rigid and death-dealing principles of Maoism or Trotskyism.

Thus the entire evolution of a certain libertarian and democratic Left, expressed in this new ethics of authenticity, supports the concern for the environment. The reformist ecologist who breaks with the radicalism of the Green parties[8] moves easily away from a call for *self-management* pure and simple, the folly of which has finally been established, to the more concrete concern with obtaining increased autonomy on *local* decisions. He demands, for example, the expansion of popular initiative referendums. A former revolutionary who readily played "society against the State," he has now become far less intransigent; like the rest of the Left, he has become pragmatic and realistic, especially if, like Brice Lalonde,[9] he has accepted to participate in the exercise of power, with its inevitable constraints. As for militants, the gentleness of democratic passions has also had an influence: commitment is now à la carte, put on display at demonstrations that with few exceptions aim to be nonviolent, using nothing but humor and irony as weapons. It is true that as in years past it is a matter of "alternative lifestyles," of "changing one's life," but the expressions no longer refer to revolution, no longer beckon to an "other world": rather they signify "living à la carte," giving oneself choices,[10] or, as the saying goes, providing a fine translation for the ethics of authenticity, "living one's life." Reconciled with the State, which gives it its own ministers, with democracy, which offers the possibility of nonviolent change, ecology ultimately blends into the market, which naturally adapts to new consumer demands. The forest is threatened by automobile emissions? No problem, we'll build catalytic converters, which are more expensive but less polluting. The docility of German manufacturers became a model: clean

[8] On this opposition within the French framework, one should read the work by Dominique Simonnet, *L'écologisme,* 3d ed. (Paris: PUF, 1992).

[9] Lalonde, it should be noted, was never really himself a "sixty-eighter."

[10] This was demonstrated in exceptional fashion by Antoine Maurice in his book, *Le surfeur et le militant* (éditions Autrement, 1987), devoted to the evolution of contemporary sensibilities in the face of ecology and sports.

industry is developing by leaps and bounds, creating competition among companies to obtain the "green" label. The supreme pardon? Perhaps. But why take offense if it allows us both to advance the cause of environmental ethics and include it within a democratic framework?

It is often asked whether ecology is a separate political force, whether it is legitimate for it to be a party unto itself with, if it comes to it, the duty to exercise power in all the traditional sectors of government. This is what the Greens would like. I think they're wrong. On an intellectual and even a philosophical level, only *deep ecology* can claim to have a global political vision—but for that it needs to hoist the flags of neoleftism or neoconservative romanticism. If ecology wishes to escape these ridiculous and dangerous archaisms, if it accepts to call itself "reformist," it will have to recognize that it is a pressure group expressing a sensibility which, though shared by the immense majority, does not have a claim to power in and of itself. As a political movement, ecology will not be democratic; as a democratic movement, it must renounce the mirage of grand political visions.

# EPILOGUE

## *Nationalism and Cosmopolitanism: The Three Cultures*

Many today seek to define Europe culturally. What do the old nations that compose it have in common, beyond the cultural divisions inherent in separate national identities? How can we look beyond our differences and together approach questions which clearly contain transnational dimensions such as those concerning the environment? In the debates that oppose the partisans of tradition with those of modernity, nationalists with the cosmopolitan, we can easily discern two ideas of a nation—an ethnic conception on the one hand and a voluntarist vision on the other.[1] The first, inherited from German romanticism, defines citizenship in terms of filiation to a race, a language, and a culture, in short, in terms of all that is *not chosen*. The second, which dates back to the French Revolution, considers a nation to be a voluntary grouping of individuals around universal principles, such as those of the Declaration of the Rights of Man. This explains the fact that we may now encounter Right-Wing and Left-Wing versions of nationalism.

The problem, of course, is that this elegant picture is a bit too pretty to be true. While flattering for the French, it happily skips over all that German thought may have contributed to the democratic ideal, and the ways in which France may have contributed to the development of fascist doctrines, not to mention colonial paternalism. Most of all, it sidesteps the paradox which makes the concept of a nation so difficult: for the two tendencies, the reference to abstract principles on the one hand and community identity on the other, are

[1] This distinction is central to the work of Louis Dumont on the birth of modern individualism.

both clearly present in the concept of the republic. Indeed, this is the most remarkable feature characterizing the modern idea of the nation: it designates both a particular cultural identity and a claim to universality. One need only read the texts devoted to this subject by our most illustrious revolutionaries: they are indissolubly French (particular) and cosmopolitan (universal), tied to a tradition and breaking away from it. Why? To cut to the main point, even if it means leaving out certain nuances, it is because the democratic idea, which was then emerging, is based on one fundamental requirement: that individuals, though culturally determined in their private spheres (the "civil society"), recognize their commonality in the principles expressed on a public level (by the representative "State"). Thus we are not forced to choose between romantic nationalism and disembodied cosmopolitanism; instead, it is on other intellectual horizons that the idea of a nation can and must find meaning today.

For it is important to understand the legitimate fear that the European construction might elicit were it to translate into an abandonment of the classical forms of politics. One theory, defended notably by the mayors of certain large cities, has it that in the future decisive questions will be handled either internationally or locally. Such a bipolar vision of politics would mean a fearsome loss of democracy, which, it is easy to foresee, will be to the benefit of demagogues.

It is true that there are questions, even outside the area of economics, which no longer make sense on a national level. Ecology falls into this category, but so do bioethics and military defense: the cloud of Chernobyl ignores national boundaries, and why forbid the trafficking of organs at home if it is legal in Luxembourg or Frankfurt? Does this mean that national politics can be emptied of all content without harm, and that the Nation-State is outdated, as certain ecologists, but also ultraliberals, argue? To believe so would incur the risk of seeing the turf abandoned by democrats immediately claimed and exploited, without competition, by the extreme Right. It is, therefore, necessary to articulate the three levels of national politics, not to renounce them in favor of a bipolarity between the local and the international. This presupposes an in-depth reflection into the role of European institutions and the ways in which they can consider the legitimate demands of national politics in a manner that is more concrete and visible for citizens.

Behind this debate, which we should all be able to judge for ourselves, a true philosophical question has reemerged: that of the status of culture in a society in which religious traditions and the transcendence of the sacred have essentially vanished. To take stock of the question, we must step outside the binary oppositions in which it often becomes ensnared: high literary culture versus technicological subculture, tradition versus modernity, romantic obscurantism versus the universalism of the Enlightenment, and so on. For many of us, the theater of contemporary culture is in effect played out within these impossible alternatives.

First, there are not two but three *philosophical* conceptions of culture constantly at odds, each aiming to supplant the other two in a fight to the death. One may, with the *utilitarians,* consider artistic works to be "products," "merchandise" which accomplishes its goal by bringing satisfaction to the public that consumes it. It is this consumerist vision that critical intellectuals denounce, justifiably in part, as the sign of a "global Americanization." Or one may, with the *romantics,* view the successful work as an expression of the particular genius of a people. Each nation possesses its "spirit," its "life," of which the language, the legal and political institutions, but also the culture in general are manifestations. These manifestations are all the more perfect as they do not throw into question that which they merely represent. Finally, at the other end of the spectrum from romanticism, one may assign cultural works the heroic task of subverting the aesthetic forms of the past, of breaking away, as the French revolutionaries saw it, from the determining codes that constitute one's national traditions. *Consumption, rootedness, separation:* these are the three key words, the three banners in the name of which the new postreligious crusades are waged. Associated with them are three possible perversions: the worship of "low culture," which is always a threat with utilitarianism, since success, the effect produced on the public, becomes the only criteria, to the exclusion of all other measures of value. The second danger is that of the descent into *nationalism and folklore,* scourge of a narrow romanticism, oscillating between fascistic arrogance and the false humility of local handicraft. As for the ideology of innovation, it also has its downside in *avant-gardism,* which has so caricatured the gesture of freedom as to reduce it to the level of bare abstraction. The content of art thus becomes reduced to a mere staging of the symbols of rupture, of subversion for subversion's sake.

The great debates on culture which we have been witnessing since the death of the avant-garde, in the mid-'70s, are often borne of the fact that these three visions of the life of the mind seem irreconcilable: avant-gardism, which dominated the "high culture" of this century, is as opposed to the romantic love of tradition as it is to the cynical reign of the market. As for romanticism, its hatred of the rootlessness inherent in modern culture is equaled only by its disdain for the plutocracy insolently displayed by the liberal world. Finally, without concerning themselves with a conflict that in their eyes opposes the elite and the obsolete, the partisans of the culture industry tranquilly pursue the production of variety shows and mercantile entertainment. Thus it is necessary to choose one's camp, and worthy intellectuals are appealing to us to separate the wheat from the chaff. Some invite us to opt in favor of innovation, to support "difficult" but "daring" culture; others recommend the safer values of national heritage and classical authors.

But are we really forced to chose between the three approaches to culture—separation, rootedness, and consumption? I don't believe so. The simplest phenomenological description of works we consider to be "great" clearly indicates that they are those which articulate these three tendencies: emancipating themselves from a context they manipulate to the point of innovation, they reconcile, each time in an original fashion, the separation and the rootedness which the avant-garde and romanticism isolate and unilaterally thematize. It is in this articulation and by it alone that they elicit the spectator's passion—the aesthetic pleasure without which even the highest culture is not worth an hour of trouble. I see no counter-example to this affirmation. It is as true for music as it is for painting, for the great mosque at Kairwan as for Notre-Dame in Paris. Both belong to a particular historical and geographical context, both transcend it, reaching a public that far surpasses that of the faithfuls to which the monument initially seemed destined. And it is in this expansion of horizons, which is impossible without the articulation of elements which some would like to separate out for ideological reasons, that true greatness resides. Here cosmopolitanism is no longer inconsistent with nationalism—even if the moment of separation from inherited codes must ultimately prevail over tradition, for without it there would be no *creation*, no innovation, and the traces of what is uniquely human would vanish. Here, it seems to me, is where the true danger lies, a danger to which we would be exposed should radical ecology succeed in winning over public opinion: by considering culture, in the manner of sociobiology, to be a simple prolongation of nature, the entire world of the mind is endangered. Between barbarism and humanism, it is now up to democratic ecology to decide.

# INDEX